# BASIC IN CHEMISTRY

a self-instructional computing course

G. BEECH

SIGMA TECHNICAL press

© G. Beech, 1976

ISBN 0 905104 00 5

All rights reserved. No part of this book may be reproduced or used in any form or by any process - mechanical, graphical or electronic (which may include information retrieval systems, photocopying or taping procedures) without the prior permission of the publisher.

Published by Sigma Technical Press,
                15a Delaware Avenue,
                Albrighton,
                West Midlands WV7 3BW.

Text prepared in IBM Bookface/Courier italic.
Printed in Great Britain at Wolverhampton Printing Co. Ltd., 6 Cleveland Road, Wolverhampton.

# BASIC IN CHEMISTRY

a self-instructional
computing course

# Contents

Preface — vii

Introduction — 1

**PART 1. SOME GENERAL PRINCIPLES**

1.1 Flow Charts for Problem Solving — 2
1.2 Computers and Languages — 5
    1.2.1 Input/Output Devices — 5
    1.2.2 Computer Languages — 6
1.3 Fundamentals of BASIC — 8
    1.3.1 Allowed Characters — 8
    1.3.2 Numerical Values — 9
    1.3.3 BASIC Program Structure — 9
    1.3.4 System Commands — 10

System Information — 11

**PART 2. BASIC STEP-BY-STEP**

UNIT 1 First Steps in BASIC — 14
    Writing and Running a Simple Program

UNIT 2 How to have Second Thoughts — 17
    Making Alterations to a Program

UNIT 3 Adding some Flexibility — 21
    READ, DATA, RESTORE and INPUT Statements

| UNIT 4 | Non-Numerical Aspects | 24 |
|---|---|---|
| | Introducing String Constants and Variables | |
| UNIT 5 | Skipping and Branching | 28 |
| | Using GOTO and IF....THEN | |
| UNIT 6 | Looping | 31 |
| | The FOR....TO and NEXT Couplet | |
| UNIT 7 | Lists and Tables of Quantities | 35 |
| | How to use Subscripted Variables | |
| UNIT 8 | Character Storage and Manipulation | 40 |
| | More about String Variables | |
| UNIT 9 | BASIC Shorthand | 45 |
| | Function Statements | |
| UNIT 10 | How to Re-Use Statements | 49 |
| | Subroutines in BASIC | |
| UNIT 11 | Elegance and Sophistication | 53 |
| | The Powerful MAT Statements | |
| UNIT 12 | Filing Useful Data | 61 |
| | Using BASIC Files to Store Instrumental Data | |

CONCLUSION TO PART 2     66

    Programming Guidelines

## PART 3. APPLICATIONS

3.1   A Survey of the Chemical Computing Literature     67

    References     71

## PART 4. RETROSPECT

4.1   Hints and Answers to Selected Problems     75

4.2   Index/MASTERFILE     78

# Preface

This book is unique in many respects. It is the first completely self-contained modular course on computer programming aimed specifically at all those involved in the physical sciences areas. By the use of examples selected mainly from chemistry and related subjects, the maximum amount of material is covered in the minimum time. Because of this approach, students at all levels of secondary and tertiary education will find that they can quickly master the fundamentals of BASIC after only a few hours of study.

The examples used in this course illustrate the various uses of computers - both as learning aids and, more simply, as powerful data processors able to solve problems beyond the scope of simple pocket calculators. Even the most involved laboratory calculations can be reduced to a simple section of BASIC programming!

I am pleased to acknowledge the cooperation of my colleagues during the preparation of this book. In particular I valued the comments of George Marr and Arnold Collett, of this Polytechnic, and John Bevan of the National Development Programme in Computer Assisted Learning. The choice of materials and the manner of presentation are however, subjective matters and comments from readers are welcomed.

Graham Beech
December 1975

Acknowledgement (Diagram, p 5)

*The author acknowledges the permission of International Computers Limited to use certain ICL material in this publication. Save for such permission all copyright, patent and other intellectual property rights belong to ICL.*

## INTRODUCTION

With the advent of small computer systems and time-sharing networks, powerful computing techniques are becoming more widely available. The newcomer is faced with a bewildering choice of computer languages, each having its own particular advantages; however, as a chemist wishing to make routine numerical calculations, you will find that BASIC has many unique advantages, not the least being its simplicity.

BASIC is an easy computer language to understand, but the best way to learn it is by reference to examples relevant to your own subject. Therefore, this course guides you through a selection of graded examples, each one introducing new concepts. The text is divided into conveniently-sized units so that you can start or stop at any point, depending on your pre-knowledge and the programming applications that you have in mind. If you are familiar with the underlying concepts of programming, you can skip most of Part 1 and read only the 'Fundamentals of BASIC' section before proceeding to Part 2. To get the best from the course, you ideally require access to a computer by a conventional teletypewriter (or similar device) although this is by no means essential. Also, you will find that you can easily complete this course in your spare time - it is completely self-contained.

Only the most fundamental features of BASIC are introduced - these being the ones commonly found on most popular machines. It is inappropriate to describe EXTENDED- or SUPER-BASIC since these versions are implemented only on specific vendor's machines. Also, most of the BASIC in this course conforms to the anticipated ANSI (American National Standards Institute) recommendations. You may find certain local peculiarities on your own machine, but these should present few problems.

The 'unit' approach of Part 2 is intended as a 'building block' system in that each unit is, to some extent, dependent on the knowledge gained from the previous units. Each unit is of roughly equal length and two or three units can be covered in a one hour session. This means that, allowing for Part 1, you could be a competent BASIC programmer within four to five hours. Clearly stated 'Performance Objectives' at the start of each unit specify the skills that you will acquire.

The scope of 'chemical computing' is illustrated in Part 3, which is an application section consisting of a fairly extensive guide to the literature of the subject. Over 80 references are cited for the interested reader.

Part 4 contains two sections: answers and hints to the problems set in Part 2 and, in place of a conventional index a learning aid which we call the MASTERFILE. This is a list of important words and phrases to which you add a brief definition or explanation for your own reference purposes.

# PART ONE

# Some General Principles

## 1.1 Flow Charts for Problem Solving

There are numerous aids to problem solving, but the general approach is to break the problem down into smaller parts and then to link these together in a logical manner. For example, a chemist may wish to analyse an aqueous solution for its iron content. He might break down this problem in the following manner:
- (1) Choose a method of analysis
- (2) Choose a standard substance
- (3) Is the standard soluble in water - if not, go back to (2)
- (4) Weigh a sample of the standard and prepare an aqueous solution
- (5) Analyse this solution by the chosen method
- (6) Compare the result with the 'true' value
- (7) If satisfactory, analyse the unknown, if not, go back to (1).

Whilst satisfactory for this simple case, such a written statement quickly becomes tedious and difficult to follow for more complex problems. A more convenient shorthand means of displaying such information graphically is to use a <u>flowchart</u>. To illustrate this, here is a flow chart based on our chemical analysis problem:

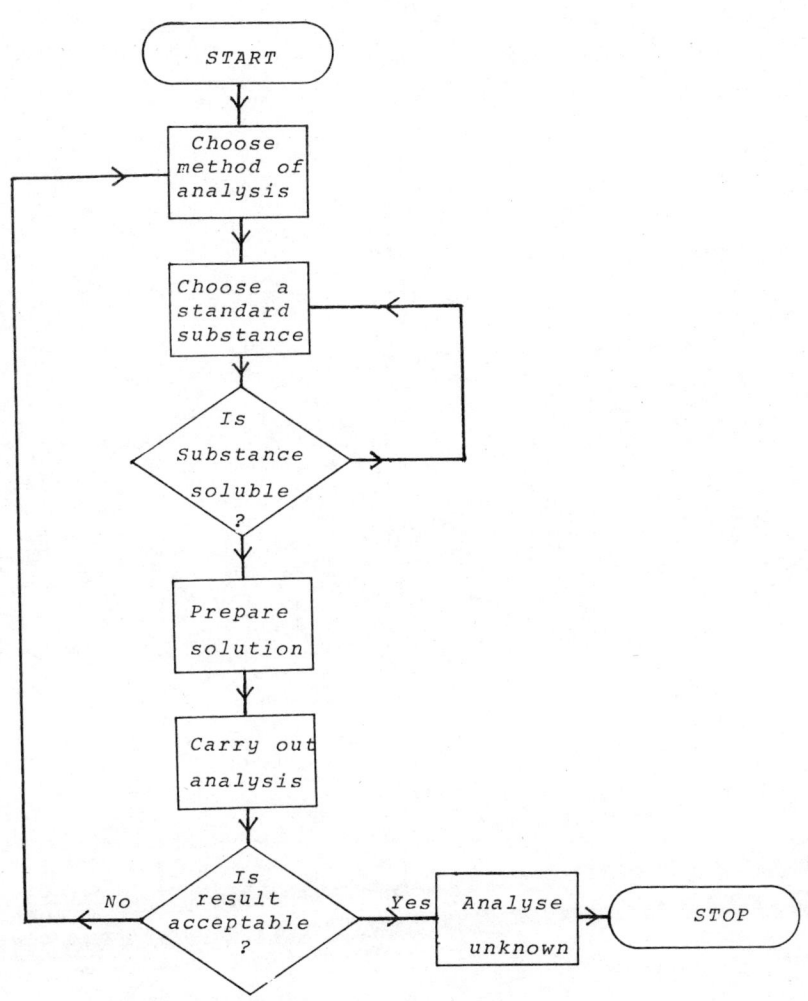

A similar method can be used to plan a computer program, which is again a sequence of logical steps, using the same symbols as in the preceding flowchart. The symbols recommended by the National Computing Centre (NCC) are:

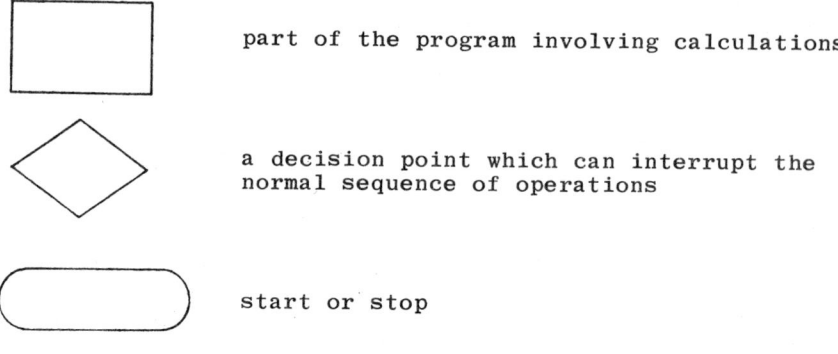

There is one more useful symbol:

input or output of data

Using only these symbols, a very simple program could be represented by the following flowchart:

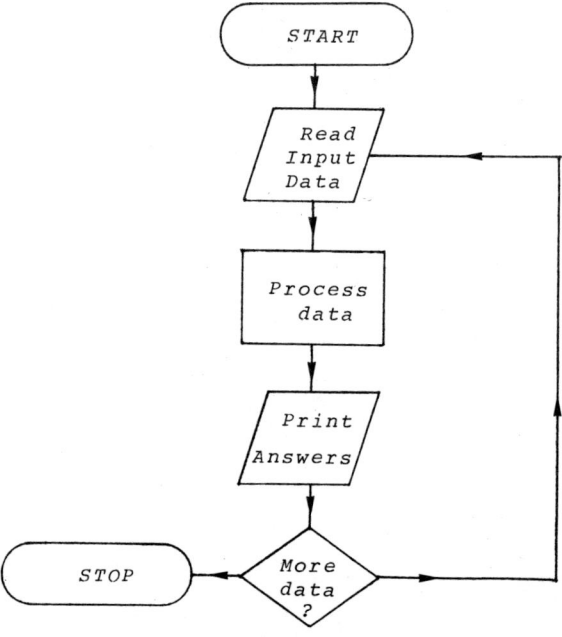

## 1.2 Computers and Languages

In this section we will limit our discussion mainly to the essentials of transmitting information to the computer with some comments on the characteristics of computer languages insofar that they affect you as a user. We also include a very brief discussion of BASIC language fundamentals as a precursor to Part 2.

### 1.2.1 Input/Output Devices

The most generally available device for BASIC programming is the teletypewriter (TTY). The TTY can be used in any one of four ways:

To transmit characters and digits to the computer by pressing the appropriate key;
To receive information from the computer and then to type it in much the same way as a conventional typewriter;
To produce punched paper tape as a permanent record of a program or data;
To read punched paper tape prepared at the same TTY or received from another source. The TTY can transmit the data on the tape to the computer.

A diagram of a typical keyboard is shown below (the one that you use may be slightly different).

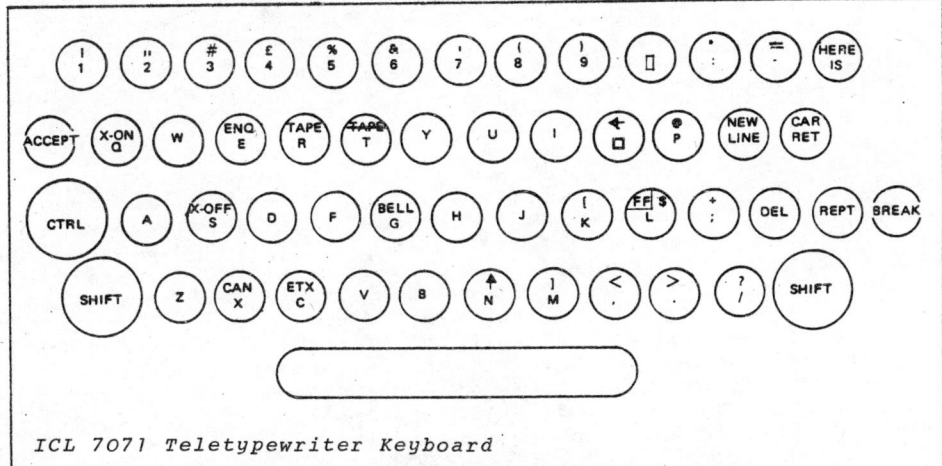

*ICL 7071 Teletypewriter Keyboard*

Not all of the keys are used but notice that some keys control more than one character. Upper register keys (e.g. $) are typed by simultaneously pressing the 'shift' key. The space character is controlled by the horizontal bar key.

Each character is encoded by the TTY into a binary representation (a series of 1's and 0's) and transmitted by wire to the computer. It is usual to communicate in units of a complete line of a program (or data) with the computer, rather than single characters, and this is achieved by hitting a special key (marked e.g. ESC, RETURN or ACCEPT) at the end of a line.

With a small computer there may be just one TTY (single user) whilst a larger machine may support many devices (multi-access). In the latter case the user still feels that he is the only user because of the way in which the computer divides its time between the various users (time sharing mode of operation).

Although we have emphasized, and will continue to do so, the TTY as an input device it is worth mentioning that other 'TTY equivalent' devices are in common use. These use a normal TTY keyboard and a cathode ray tube display and are called visual display units, (VDU); their main advantage is their faster speed of operation although in their simplest forms they have the disadvantage of only <u>displaying</u> the information on the screen for a transient time. Both the TTY and VDU devices, known generically as <u>terminals</u>, also allow information transmitted <u>from</u> the computer to be displayed. This enables the user to establish an interactive dialogue with the computer which is all important in BASIC programming, since it permits the rapid detection of syntax errors (these arise when the grammatical rules of the language are not obeyed).

Another long-established, but non-interactive, means of getting programs and data into a computer is by punched cards or punched tape. The latter can often be prepared at the TTY (although certain vendor's machines have somewhat eccentric conventions for paper tape input) and they are particularly useful if a program is to be used elsewhere or if the same data will be needed at a later date.

## 1.2.2 Computer Languages

The operations of a computer are controlled by a sequence of simple instructions. This sequence, which we recognise as a computer program, is stored in the main computer memory (in the case of a time-sharing machine it is stored only when actually carrying out useful operations, otherwise it is stored in auxilliary memory such as magnetic tape or disc).

The information-processing heart of the computer can not, however, understand the near-English programming languages that we would prefer to use. The only instructions that can be directly used by our computer are 'machine code' instructions. Each such instruction is a sequence of 1's and 0's, for example:

$$0101100000000001$$

Each of the above sixteen binary digits is called a bit; bits may be grouped into <u>bytes</u> (always of 8 bits) or a <u>word</u> (usually more than 8 bits). The example above is a word of 16 bits, this being a popular word length on small machines, although other lengths will be found. However, even though our simple-minded computer readily understands these binary commands, only the most dedicated human being would try to use this outlandish 'language'.

Fortunately, programs <u>can</u> be written in near-English dialects which bridge the gap between the machine-oriented instructions, described above, and the type of mathematical representation familiar to you as a scientist. This is

achieved by so-called 'high-level' languages which can be
easily understood by the user.  A computer program in such
a language consists of a series of easily recognisable
statements, such as:
    90 LET X=(Y+Z)/S      which means 'set x equal to (y+z)÷s'
    100 LET A= B+C/D +E↑2            'set a equal to b+(c÷d)+e$^2$'
Each statement is translated into 'machine code' binary
instructions by a special program stored in the computer
called a "compiler" and each high level statement may, in fact,
generate more than one machine code ('low level') instruction.
This is illustrated in the following diagram:

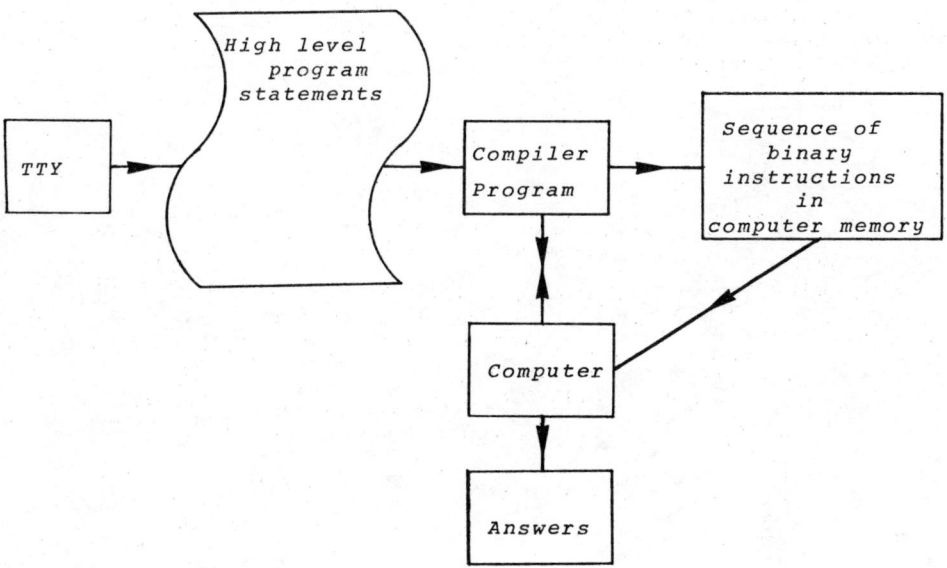

    Each high level language requires a separate compiler;
some languages, including most interactive variants of BASIC,
have compilers which check the program line by line and the
user is then provided with instant diagnostics relating to
syntax errors in his program statements.
    The most important high level languages for the scientific
user are
                  BASIC, FORTRAN and ALGOL.
Each of these permits you to write programs in a nearly-
conventional scientific form, using familiar symbols and
notation.  The main advantage of BASIC, besides simplicity,
is that if the program is being typed in an interactive mode,

then each line is checked by the compiler as it is entered to see that it conforms to the grammatical rules ('syntax') of the BASIC language.  For example, if we type
                    100 LET A EQUALS B+C
the computer might respond with a message such as
                         ?100 LET A/
because it does not recognise the word EQUAL.  Different computer systems have many ways of presenting this diagnostic information but, clearly, this dynamic interaction with the user is an important feature of the language.  TTY's and VDU's also permit the user to erase individual characters or to make other alterations.

If the complete program has been typed, free of syntax errors, we instruct the computer to run the program - i.e. to obey the program instructions in strictly-defined order and to use or request data.  At this stage other errors may be automatically detected and reported to the user.  These will be of a logical nature; for example, attempting to divide by zero or to take the square root of a negative number.  Some systems allow the user to work with BASIC in a "calculator mode", which can also be helpful at this diagnostic stage.

Having set the scene we will now examine the fundamentals of the BASIC language prior to a more detailed study.

## 1.3   Fundamentals of BASIC

A BASIC program consists of a series of numbered lines or statements.  Each statement is a specific instruction to carry out a particular operation.  The most useful of these are for:
> simple arithmetic
> reading the input data
> printing the results
> comparison tests

### 1.3.1 Allowed Characters
Each statement is a combination of certain allowed letters, numbers and other symbols which will be defined here for convenience:
Letters:   any capital letter from A to Z, being careful to distinguish:
         Z and 2        S and 5        O (letter 'oh') and 0 (zero).
For clarity, you can write Ø for zero where ambiguity is possible.
Numbers:   Ø to 9 inclusive.
Special characters:  be sure to learn these or you will not be able to follow the rest of the course.  The most commonly used characters are:
- `+`  addition or positive sign
- `=`  equals
- `*`  multiplication
- `()` brackets
- `,`  comma
- `-`  subtraction or negative sign
- `/`  division
- `↑`  raising to a power
- `.`  decimal point

Other useful characters include:
```
  > greater than              < less than
  ; semi colon                : colon
  " quote mark                $ or £ dollar or pound sign
  # hash mark                        (often used interchangeably)
```
Certain combinations of these characters are also permitted.

### 1.3.2 Numerical Values

These can be used in BASIC programs but there is a limit to the number of digits that are used to represent a number in the normal fashion. This varies between 6 and 15 digits, depending upon the particular vendor; so, if for example, 8 digits are permitted then the following are all valid numbers:

```
        3.82           237826.01      9.832
       -1579            -38234        0.0237899
```

The following would not be permitted:

```
       19382.43798   (more than 8 digits)
       19,382        (commas are not permitted)
```

Numbers with more than the permitted number of digits are represented in the form:

$$aEb$$

where E means '10 raised to the power'. In other words we multiply 'a' by 10 raised to the power 'b'. For example:

```
    22467800000  becomes  2.24678E10  or  224.678E8 etc.
          327.4  becomes  3.274E2     or  32.74E1 or 3274.0E-1 etc.
     0.0000000123 becomes 1.23E-8     or  0.00123E-5 etc.
```

In this notation we can enter (on the TTY) the largest and smallest numbers that can be stored in the computer. (For example, the type of BASIC used on our ICL computer permits numbers in the approximate range of $10^{-75}$ to $10^{+75}$. The computer will also output very large or small numbers in this form on the TTY.

### 1.3.3 BASIC Program Structure

Each line of a BASIC program has a number or 'label' followed by a BASIC statement. An example is:

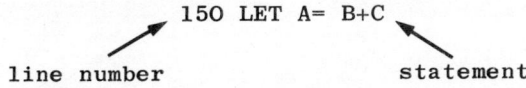

150 LET A= B+C

      line number            statement

Statements are normally obeyed in ascending order of line number, although it is worth remembering that they can be <u>typed</u> in any order of line number - sorting into ascending order is taken care of by the computer. We normally number the lines in steps of 10 because this enables us to later insert extra lines without extensive alteration. For example, we may have written a program with lines numbered from 100 to 650 in steps of 10. Before running the program we realise that an extra statement must be inserted between lines 140 and 150. Therefore we could type:

$$145 \quad new\ statement$$

and then run the program. This is an example of one of the simplest BASIC statements called an "assignment" statement. In this case, it takes two numerical quantities represented by B and C and *assigns* their sum to A. These quantities A, B and C are called "numerical variables". Therefore, our statement is <u>not</u> an equation, because we could also write:

$$170\ \text{LET}\ A = A+2$$

and A does <u>not</u> cancel on each side of the equals sign.
BASIC is also very tolerant of eccentricities in typing. Consider these examples:
    (i) a conventional layout:
$$120\ \text{LET}\ A1 = A + B$$
    (ii) the less well-typed, but equally acceptable form:
$$120 \quad \text{LE} \quad\quad \text{TA} \quad\quad 1= \quad\quad A+ \quad\quad B$$
Although the spacing is quite different, this would have precisely the same effect as the statement in (i). Therefore, your typing skill has no influence on the final result, although we will naturally prefer the neater layout.

### 1.3.4 System Commands

The user has to issue commands to the computer to inform it, for example, that he intends to write a BASIC program or that, having written it, he wants to run it. These are examples of 'system commands' and they are <u>not</u> a part of the BASIC language. For this reason, there is considerable variation between the commands used on different vendor's machines. Nevertheless, there is some similarity in the <u>sequence</u> of commands, of which the following is typical:

<u>Activate the TTY</u> - either a dial-up procedure or a special combination of keys. From hereon we wait for the 'prompt' character, (such as ?, ←, -, * etc.) before typing a line and end each line by pressing the terminator key (e.g. ESC, RETURN or ACCEPT).

<u>Log-on</u> - most computer systems require the user to identify himself with his user name or code.

<u>Request compiler</u> - often just by typing BASIC.

<u>Computer types NEW or OLD?</u> - if the program is being written for the first time we type NEW followed by our chosen program name. Then type the program:

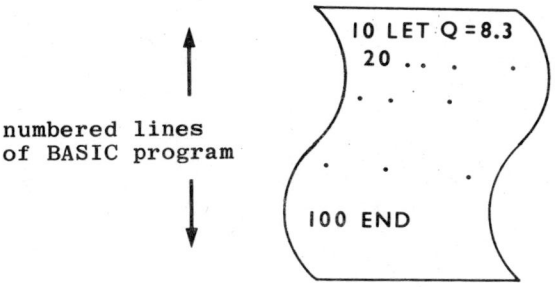

numbered lines
of BASIC program

<u>Run the program</u> — by typing RUN, the statements are obeyed in line-number order
<u>Terminate the session</u> — type BYE or GOODBYE
<u>Log off</u> — usually a special character sequence after which the computer may inform you of your 'connect time' and other statistics.

As this sequence depends on the particular computer you are using, the following section will be useful to you:

---

**System Information**

You will need to know the various system commands in operation on your particular computer. It will help you considerably if the list of requirements on the next page is completed by a member of your computer staff. The information should also be available in the appropriate reference manuals.

*The student who is using this book will want to know the various system commands in operation on his local computer installation. Please will you tell him:*
*(i) how to activate the TTY...................................*
*(ii) what logging-on procedure is used (he may need a user number)......................................................*
*(iii) how to request the BASIC compiler.......................*
*..............................................................*
*(iv) what to type in the case of a new BASIC program (this may be the word NEW-compare ICL X98B BASIC).......................*
*..............................................................*
*(v) how to correct individual characters and how to erase complete or partial lines of input; if VDU's are used, he also needs to know about the cursor......................................*
*..............................................................*
*(vi) what command is used to cause a BASIC program to run*
*..............................................................*
*(vii) if BASIC in 'calculator mode' is available, please explain briefly..............................................*
*..............................................................*
*(viii) how is the BASIC session terminated....................*
*..............................................................*
*(ix) is there a log-off procedure.............................*

*Later in the course (UNITS 2-11) he will also want to know:*
*(x) how to store (save) a BASIC program......................*
*..............................................................*
*(xi) how to use a previously-stored BASIC program.............*
*..............................................................*
*(xii) how to erase a previously-stored BASIC program*
*..............................................................*
*(xiii) if there is a LIST command.............................*
*(xiv) if (UNIT 8) it is necessary to declare the length of a character string in a DIM statement or if it assumed that a maximum number of characters can be stored in a string variable (please specify)............................................*
*..............................................................*
*..............................................................*
*(xv) are MAT statements (UNIT 11) permitted...................*
*..............................................................*

*If he intends to use UNIT 12, he will also need to know the local rules for BASIC file handling. The ones in UNIT 12 are those available on the ICL X98B BASIC system...............*
*..............................................................*
*..............................................................*

# PART TWO

# BASIC Step-by-Step

## UNIT 1.   FIRST STEPS IN BASIC

*Performance Objectives:   At the end of this unit you should be able to:*
  *(1) Make simple arithmetical calculations, using BASIC statements;*
  *(2) Assemble these, with suitable print statements and the simplest system commands, into a useful program.*

A useful BASIC program can be written with only a few simple statements and commands.  Before typing your program, however, the BASIC compiler will need to know if you are creating a new program or using an old one.  Many systems require you to start on the first line of a program with the words NEW or OLD respectively followed by the name given to the program.  Note that this is <u>not</u> part of the BASIC language and the precise implementation may be peculiar to your machine (see page 12).

We will describe three simple statements; namely, those using LET, PRINT and END.  The first two statements are used to perform calculations and to print results.  The calculations may involve purely numerical quantities (e.g. 2.023) or symbols (e.g. X , Y) to represent numbers.  The former are called "numerical constants" and the latter are "numerical variables".  They can be positive or negative and their magnitudes are constrained solely by the design of the computer.  Constants can be expressed in either the conventional or E formats (page 9).  Variables are represented by a single capital letter from A to Z or by a single capital letter followed by a single non-zero digit.  The three statements are described as follows:

| Statement | Example | Explanation |
|---|---|---|
| (assignment see p.10) | 100 LET Y=4+1/2<br>120 LET X=(4+1)/2 | Y and X are given the values $4\frac{1}{2}$ and $2\frac{1}{2}$ respectively |
| | 130 LET X=X+2 | X is <u>assigned</u> its previous value ($4\frac{1}{2}$) plus 2.  Note that X does <u>not</u> cancel on each side of the equals sign. |
| | 190 LET Q2=Y+X1-C4<br>200 LET R6=(X↑2-Y)/(4+A) | Variables are represented by a capital letter on its own <u>or</u> followed by a single non-zero digit. |
| PRINT | 250 PRINT X<br>300 PRINT X,Y | <u>6.5</u><br><u>6.5</u>          <u>4.5</u><br>(We are underlining TTY responses for clarity). |

| | | |
|---|---|---|
| END | 900 END | This must be present to mark the end of the program |

Make a particularly careful note of the use of brackets; A common error is to forget to use brackets in the denominator of an expression. For example, the ratio 4/2*2 would be calculated, in BASIC, to be 4 i.e. (4/2*4 whereas 4/(2*2) is, as expected, equal to 1. Also, no two operators can be adjacent so that, for example, this statement is illegal:

100 LET X=4*-2

This must be changed to

100 LET X=4*(-2)

so that brackets must be used to avoid this type of error.

A program comprised of a complete series of statements is followed by the command RUN and each numbered statement will then be obeyed in order of ascending line number.

---

*Example Problem: Program to Calculate Atomic Masses*

This program calculates the atomic mass of chlorine from the unknown masses of the chlorine isotopes. The masses are 34.96885 (75.53%) and 36.96590 (24.47%). The formula used is:

Atomic mass = (34.96885 x 75.53 + 36.96590 x 24.47)/100

Any other combination of masses and abundance could have been used. A possible flow chart is shown overleaf with the corresponding BASIC program. The numerical value after the word RUN is the computed atomic mass.

Thus the computed atomic mass of chlorine is 35.4575. You will have noticed that this program was not designed for efficient operation - lines 100 to 150 could have been combined into one line - but it does enable you to change A1, A2, P1 and P2.

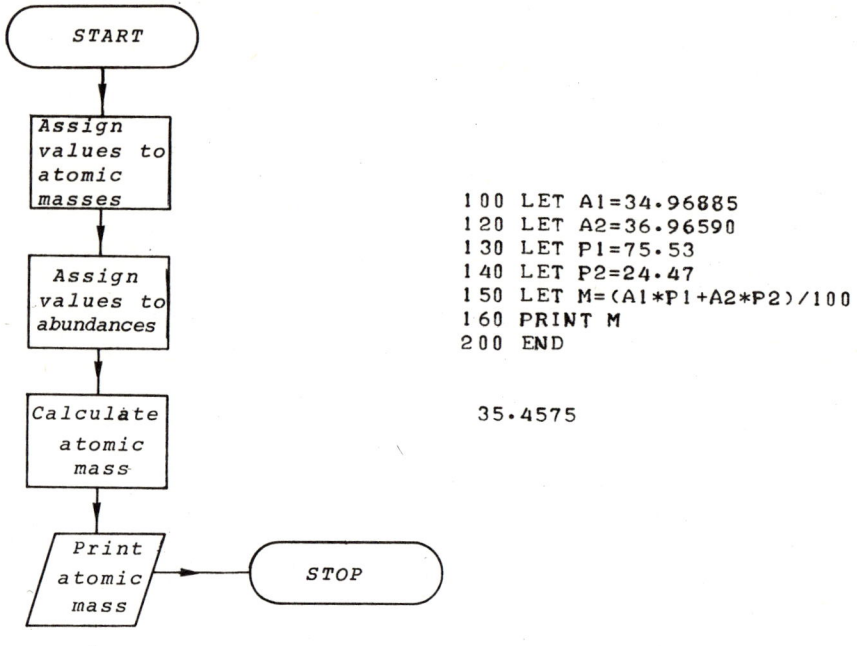

```
100 LET A1=34.96885
120 LET A2=36.96590
130 LET P1=75.53
140 LET P2=24.47
150 LET M=(A1*P1+A2*P2)/100
160 PRINT M
200 END

35.4575
```

*Practice Problems*

(1) Correct the following BASIC statements:
```
100 LET M= X10 + K          120 LET 92=Y
520 LET R=U+P/-3            999 Y=P+Q
```

(2) Choose a different pair of isotopes and substitute the appropriate data into the program above. Re-run the modified program.

(3) In each of the following cases, convert any number in E representation to normal arithmetical notation; otherwise, change it back:

        2.78E2           -89.2
        48.376E-3      -3.2845E6
        589,488,732.43  0.0000000000372

(4) Write a program to calculate the volume, V, in litres of an ideal gas from the equation

$$PV = nRT$$

where P = pressure in atmospheres
      R = gas constant (0.0820552 litre atm $K^{-1}$ $mol^{-1}$)
      T = absolute temperature (in degrees K)

## UNIT 2. HOW TO HAVE SECOND THOUGHTS

*Performance Objectives: At the end of this unit, and after referring to the specific instructions for your computer, you should be able to:*
- *(1) Delete lines from a program;*
- *(2) Insert lines into a program;*
- *(3) Use other common system commands for your particular computer.*

On page 8, we mentioned the ways in which errors were detected. Most BASIC systems also allow you to replace, delete or add lines in a current program before terminating the session. In this way, the program in UNIT 1 could be easily adapted for other combinations of isotopes by retyping lines 100 to 140. You would simply type the number of the line to be altered, followed by the new statement. For example, if you had just typed the program in UNIT 1, and wished to set the variable A1 to a different value, you might type

```
100 LET A1=22.1483
```

This would replace the previous line numbered 100 and it could be typed at any time in the BASIC session - before or after the original program had been run. Therefore, the following two BASIC sessions yield equivalent results after the final RUN:

Session 1

```
100 LET A1=34.96885
120 LET A2=36.9659
100 LET A1=22.1483
130 LET P1=75.53
140 LET P2=24.47
150 LET M=(A1*P1+A2*P2)/100
160 PRINT M
200 END
RUN
```

25.7742

Session 2

```
100 LET A1=34.96885
120 LET A2=36.9659
130 LET P1=75.53
140 LET P2=24.47
150 LET M=(A1*P1+A2*P2)/100
160 PRINT M
200 END
RUN
```

35.4575

```
100 LET A1=22.1483
RUN
```

25.7742

(TTY responses are underlined)

Equally easily, lines can be <u>inserted between</u> existing ones so long as the line numbers permit it. For example, we might type

```
125 PRINT A1, A2
```

This would be automatically inserted between lines 120 and 130 in the above program. A line can be <u>deleted</u> by typing just its line number; for example, if you type

```
125
```

then the line number 125 would no longer exist.

If the user now types SAVE, (this is the ICL command for your system, see page 12), the program will be copied by the computer system from the 'work file' which is a temporary storage area in main memory allocated by the computer for the development of this specific program, into a permanent file which may be stored in auxiliary memory such as magnetic disc or tape.  On returning to the terminal at some future time and going through the sequence on page 10, we will again receive the NEW or OLD request, (again this is for ICL and certain other machines).  If we now type

OLD *program*

the BASIC system will find the previously SAVE'd program and you can then run it or modify it, just as you would with a NEW program.

Although, as mentioned previously (pages 10-12), there is considerable variation in the precise implementation of these <u>system commands</u>, it is worthwhile to look in somewhat more detail at a typical interactive dialogue on two different computer systems.  Following this, we will list some of the more useful commands with which you must familiarise yourself.

*Example 1 : dialogue with a Hewlett-Packard 2000E time-shared BASIC system*

In this case, the TTY is linked to a central computer by a conventional telephone line.  When a connection has been made between the TTY and the computer, the user "logs-on" and might then type (<u>TTY responses are underlined</u>):

```
HEL-X123, code              "code" is some allocated password
READY
100 ......  ⎤
110 ......  ⎥             BASIC program
..........  ⎥
999 END     ⎦
RUN
......      ⎤             Results
......      ⎦
DONE
NAM-PROG1                   Program is given the name PROG1
SAV                         A copy is stored on disc
NAM-PROG2                   A new program PROG2 is to be typed
100......   ⎤
110......   ⎥             BASIC program
..........  ⎥
1000 END    ⎦
RUN
......
......
DONE
SAV
GET-PROG1                   User requires the program which was
                            previously SAVed
RUN
```

```
PROG1
......  ⎤
......  ⎦                     Results
DONE
100 statement                 line 100 in PROG1 is altered
KIL-PROG1                     PROG1 is erased from disc
SAV                           altered version is SAVed
BYE                           session terminated
```

*Example 2 : dialogue with an ICL 1900 series time sharing system (operating under the GEORGE III operating system)*

A special combination of TTY characters is used to activate the terminal, to which the system responds with information of the type in example 1; after 'logging in' with his code number the user could have this dialogue, most of which is self-explanatory on comparison with example 1 (TTY output is, again, underlined):

```
TYPE PASSWORD  ANALYST        for file security, if needed
BASIC
NEW OR OLD? NEW PROG1         gives a name to the program
......  ⎤
......  ⎥
......  ⎥                     BASIC statements of PROG1
......  ⎥
......  ⎥
......  ⎦
RUN
SAVE                          equivalent to SAV
OLD PROG2                     equivalent to GET
PROG2 IS BEING RETRIEVED
LIST
......  ⎤
......  ⎥
......  ⎥                     listing of PROG2 produced
......  ⎥
......  ⎦
RUN
UNSAVE PROG2                  equivalent to KIL
BYE
```

The user then 'logs out' after which plentiful, but unimportant data is typed by the system.

As you can see, the system commands are similar but not identical. Therefore, it is imperative that you familiarise yourself with the common system commands for your local BASIC system. In addition to the logging on or off procedures, you should find the equivalents of each of the commands used in example 1 or 2. These should have been noted at the end of section 1.3.4 in Part 1 of this book. Make sure that you can use the commands before proceeding to UNIT 3. Since we will need a convention, the commands in <u>example 2</u> (those for the ICL 1900 series) will be used in the rest of the course.

*Practice Problems*

(1) Re-type the program in UNIT 1 and repeat exercise 1 in the unit. First, RUN the original program then make the amendments in the current session.
(2) SAVE the program that you have just written and amended and terminate the session. Commence a new session, retrieve the program and re-run it.

## UNIT 3. ADDING SOME FLEXIBILITY

*Performance Objectives:* At the end of this unit you should be able to:
(1) Use the READ/DATA couplet to assign values to constants;
(2) Use the INPUT statement to assign values to variables.

Some additional features are now introduced which avoid the need to alter lines in a program each time a different set of data is to be input:

| Statement | Purpose | Example | Explanation |
|---|---|---|---|
| READ | Assign constant values to variables from a list of data | 230 READ X1<br>340 READ A,B | Each READ must have a corresponding DATA statement (below) |
| DATA | Creates a list of data items which must be separated by commas if there is more than 1 item | 550 DATA 2.3<br><br>620 DATA 4,9 | Assigns 2.3 to X1 in statement 230<br><br>These values are assigned to A and B in statement 340 |

READ and DATA are normally used for the storage of constants within a program. The numerical values in a DATA list are accessed sequently by the READ statements but, occasionally, the <u>same</u> list of values in a DATA statement will need to be assigned to two or more <u>different</u> sets of variables. This can be achieved with the RESTORE statement to re-establish the original list. For example:

```
110 READ A,B,C
110 RESTORE
120 READ P,Q,R
 . . . . . . . .
 . . . . . . . .
900 DATA 500.32,-45.69,28.3
```

Then A and P will have the same values, likewise B and Q, and C and R. However, you should be very careful in the use of READ/DATA because all of the constants in the DATA statements are sorted into one "stack" to which all READ's refer. The RESTORE re-establishes the complete stack sequence - not just a part of it.

If we wish to use different values of specified variables in subsequent runs of a program, then INPUT is preferred.

| Statement | Purpose | Example | Explanation |
|---|---|---|---|
| INPUT | Requests values to be assigned to variables from the TTY when the INPUT statement is encountered at RUN time. | 230 INPUT X1<br><br>270 INPUT A,B,C | User could type 2.3<br><br>User could type 3.5, 9.97, -4.2 |

The final statement introduced in this section is REM, which is used simply to add lines of explanatory comments (i.e. REMarks):
>    200 REM - THIS IS A BASIC PROGRAM

We will now use each of these statements in modifications of ATOMICMASS (UNIT 1).

---

*Example Problems*

 Our first version, suitable for interactive or non-interactive computing, uses the READ and DATA statements

```
 90 REM-VERSION 1
100 READ A1,A2,P1,P2
110 LET M=(A1*P1+A2*P2)/100
120 PRINT M
130 DATA 34.96885,36.9659,75.53,24.47
140 END

    RUN

 35.4575
```

Notice how this has shortened the program <u>and</u> added flexibility since only line 130 needs to be altered for a different set of data. Note also that DATA statements can be put anywhere but are convenient to group together at the end of a program.

 An interactive version for use at a teletype is even shorter:

```
 90 REM-VERSION 2
100 INPUT A1,A2,P1,P2
110 LET M=(A1*P1+A2*P2)/100
120 PRINT M
130 END
```

When line 100 is reached the computer "prompts" the person at the TTY to input the four items of data (either a message such as "INPUT 4" or a special character is typed). Therefore, at such an invitation we would type

>        34.96885,36.9659,75.53,24.47

and the TTY would give the same numerical answer as before. Note that commas are used to separate the items, just as if they were in a DATA statement.

 As you can see, INPUT is more flexible than READ and is more suitable when an interactive "coversation" is needed. READ is useful when changes in data are infrequent or when the program and data are being prepared on paper tape for non-interactive use.

*Practice Problems*

(1) Correct the following BASIC statements:
       (i) 100 READ A,B,C,D   (ii) DATA 8 9 10
           999 DATA 4,5,6
     (iii) 800 INPUT 8 t,Y

(2) Write a program to calculate the frequencies in wavenumbers ($cm^{-1}$) of the emission lines in the spectrum of atomic hydrogen using the formula:
$$\bar{\nu} = R(1/n_1^2 - 1/n_2^2)$$
when R (the Rydberg constant) is equal to 109 677.581 $cm^{-1}$
The integers $n_1$ and $n_2$ are numbers ('quantum numbers') which relate to the lower and upper energy states of the hydrogen atom (e.g. $n_1$ could be 2 and then $n_2$ could be 3,4,5 etc.). Use READ or INPUT to assign the values of $n_1$ and $n_2$. Compare your results with values in the literature ('Physical Chemistry' by W.J.Moore 4th Edn. Longman, London, 1963, p470-2).

(3) Try to write a program to calculate the mole fractions of naphthalene and p-dichlorobenzene in a mixture of the two. If we use the abbreviations

       W1 = number of grams of naphthalene
       M1 = molar mass of naphthalene
       W2 = number of grams of p-dichlorobenzene
       M2 = molar mass of dichlorobenzene

then the corresponding mole fractions are:

       X1 = (W1/M1)/(W1/M1 + W2/M2)
       X2 = (W2/M2)/(W1/M1 + W2/M2)

You can store M1 and M2 in a DATA statement and INPUT the values of W1 and W2. Any other combination of two or more compounds could be used, with appropriate modifications. (For example, the values of the molar masses in the DATA statement could easily be changed).

## UNIT 4. NON-NUMERICAL ASPECTS

*Performance Objectives: At the end of this unit, you should be able to:*
  *(1) Use the PRINT statement to control the layout of printed results;*
  *(2) Enhance printed output with simple messages to the terminal user.*

---

As you have seen, results from a program can be displayed with the PRINT statement. If several items in a single PRINT are separated by commas, then the items will be spaced 15 horizontal spaces apart measured from the beginning of each number. Usually, 6 to 9 digits are used to print numerical answers excluding surplus zeroes; other numbers are represented in the E notation (page 9). For example:

```
100 LET X = -95.300000
120 LET Y = (2E10)/0.25
130 PRINT X,Y,-10/3
```

would give

```
-95.3            8E10            -3.3333333
```

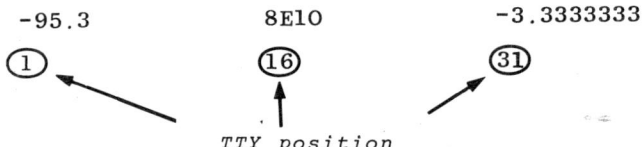

*TTY position*

If the items in a print list are separated by semi-colons the output is more compact, with only 1 to 3 spaces between each item:

`130 PRINT X;Y;-10/3`

would give

`-95.3 8E10 -3.3333333`

Even more control is possible with the TAB(x) statement which simply moves the TTY print head to position x. For example:

`130 PRINT TAB(10);X;Y;TAB(52);-10/3`

would give

```
          -95.3 8E10                              -3.3333333
```

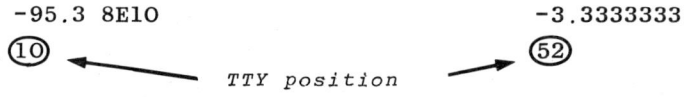

*TTY position*

Informative words or phrases can be included in PRINT statements and this is almost essential in interactive programs of any complexity. Such data goes under the general name of "character strings" and must be enclosed in quote marks - for example:

`130 PRINT"THE VALUES ARE";X;Y;"AND";-10/3`

gives us

`THE VALUES ARE -95.3 8E10 AND 3.3333333`

Words and phrases can also be <u>stored</u> in <u>string variables</u> which are denoted by a single letter followed by a dollar sign, $. The word or phrase should not exceed 15 characters (including spaces) on many computers; some other computers require that you specify the number of characters in the string with a DIM statement (UNIT 7) and this practice is likely to become more widespread. An example of the use of string variables is:

100 PRINT"TYPE A MESSAGE";
110 INPUT A$
120 PRINT"MESSAGE TYPED IS";A$

So, if in response to line 100, you typed BASIC IS EASY then the TTY would type

MESSAGE TYPED IS BASIC IS EASY

Much more information on character strings is to be found in UNITS 5 and 8.
Finally, you can skip a line with a blank PRINT statement:
180 PRINT
This can also be used to space to a new line if the last item in the previous PRINT list was followed by a comma or semi-colon (this happens in UNIT 8) otherwise the TTY will carry on printing on the <u>same</u> line.

---

*Example Problem - Program to Calculate Radioactivity after Neutron Activation*
    The radioactivity of a substance after irradiation by neutrons (e.g. in a nuclear reactor) after a time T is given by

$$A = F \cdot S \cdot N (1 - 0.5^{T/t_{\frac{1}{2}}})$$

where $t_{\frac{1}{2}}$ is the half life of the product in the same time units as the irradiation time T.
If  F=neutron flux, in neutrons $cm^{-2} s^{-1}$
    S=nuclear reaction cross section, in $cm^2$
and N=number of nuclei in the sample
then
    A=activity of the sample in <u>disintegrations per second</u> (dps)
We obtain a more useful equation by expressing S in barns
(1 barn =$10^{-24} cm^{-2}$) and calculating N in terms of the mass, m, of the sample and its atomic mass, W:

$$A = \frac{0.6 \; F \cdot S \cdot m}{3.7 \times 10^{10} \; W} (1 - 0.5^{T/t_{\frac{1}{2}}})$$

The activity, A, is now in "Curies" (1 Curie =$3.7 \times 10^{10}$ dps) which are the accepted units of radioactivity.

A flowchart to compute values of activities is shown overleaf.

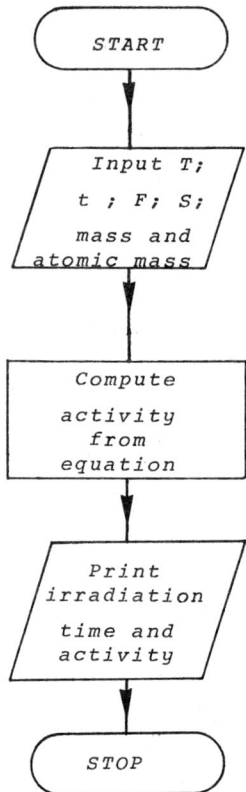

The following program uses this flowchart and makes use of the BASIC statements introduced in this unit:

```
90 REM-PROGRAM TO CALCULATE RADIOACTIVITY AFTER NEUTRON IRRADIATION
100 PRINT"INPUT IRRADIATION TIME AND HALF LIFE IN THE SAME UNITS"
105 PRINT"SEPARATED BY A COMMA";
110 INPUT T,T1
120 PRINT"INPUT FLUX VALUE";
130 INPUT F
140 PRINT"INPUT CROSS SECTION IN BARNS";
150 INPUT S
160 PRINT"INPUT MASS OF SAMPLE IN GRAMS";
170 INPUT M
180 PRINT"INPUT ATOMIC MASS OF ELEMENT";
190 INPUT W
195 REM-NOTE THE USE OF E IN THE NEXT LINE
200 LET A=(0.6*F*S*M*(1-0.5↑(T/T1)))/((3.7E10)*W)
210 PRINT"ACTIVITY AT TIME";T;"IS";A;"CURIES"
220 END
```

```
RUN

INPUT IRRADIATION TIME AND HALF LIFE IN THE SAME UNITS
SEPARATED BY A COMMA← 5,2
INPUT FLUX VALUE← 1.2E12
INPUT CROSS SECTION IN BARNS← 3
INPUT MASS OF SAMPLE IN GRAMS← 0.5
INPUT ATOMIC MASS OF ELEMENT← 62
ACTIVITY AT TIME 5 IS .387568 CURIES
```

You can see how the output from the program and its interactive character have benefited from the use of character strings. We will make further use of these strings in the next section.

*Practice Problems*

1) Correct the following BASIC statements:

   1900 PRINT 'MESSAGE'          450 PRINT "A$"
   900 PRINT Q;LAST VALUE

2) Be sure that you can modify the program ACTIVATE in this unit so that it uses READ and DATA in place of INPUT.

3) Write a short program to calculate and print the sulphur and chlorine contents of a sample so that the results are set out as follows:

   PERCENT SULPHUR=          .CHLORINE=

   Assume that input to the program consists of the weights of sulphur and chlorine, x and y respectively, and z, the total weight of the sample. (As an example, with x=2.14, y=0.83 and z=5.5 the answers should be 38.91 and 15.09 respectively).

4) If an ideal gas is forced to expand adiabatically and reversibly from an initial pressure $P_1$ to a final $P_2$, then its final temperature $T_2$ is given by

$$T_2 = \frac{P_2}{nR}\left(\frac{P_1}{P_2}\right)^{\gamma} V_1$$

where

$V_1$ = initial volume in litres
$n$ = number of moles of the gas
$R$ = the gas constant (0.0820552 litre atm $K^{-1}$ $mol^{-1}$)
$\gamma$ = the heat capacity ratio, $C_p/C_v$ (5/3 for a monatomic ideal gas)

Construct your program so that it will instruct the user to input any combination of $P_1$, $P_2$, $V_1$ or $\gamma$.

## UNIT 5.  SKIPPING AND BRANCHING

*Performance Objectives: At the end of this unit you should be able to:*
  *(1) Make an unconditional transfer with a GO TO statement;*
  *(2) Make conditional transfers with IF....THEN statements.*
  *(3) Use string variables to control the logical flow of a program.*

You will recall that, in a BASIC program, the statements are normally obeyed in order of increasing line number (p.9). On many occasions, however, the program <u>branches from its normal sequence</u> if some pre-arranged condition is satisfied. We have touched on this point in our discussion of flowcharts (p.4).  There are two types of BASIC branching commands - unconditional and conditional ones.

The former type are ones that are obeyed regardless of the state of any variables in the program.  An example of such a statement is GOTO which causes the computer to automatically and unconditionally skip to a specified line.  For example,

$$180 \text{ GOTO } 250$$

transfers control to line 250.
We can also transfer control subject to some selected condition such as

$$180 \text{ IF } Y=20 \text{ THEN } 250$$

The IF....THEN combination only transfers control to line 250 in this case when the condition Y=20 is true.  Other conditions are also possible:

| Condition | Example | Symbol |
|---|---|---|
| greater than | 180 IF Y > 20 THEN 250 | > |
| less than | 180 IF Y < 20 THEN 250 | < |
| not greater than | 180 IF Y <= 20 THEN 250 | <= |
| not less than | 180 IF Y >= 20 THEN 250 | >= |
| not equal to | 180 IF Y <> 20 THEN 250 | <> |

Any variables, including strings can be used in such comparisons:

```
2000 PRINT"WHAT UNIT IS THIS?";
2010 INPUT A$
2020 IF A$="UNIT 5" THEN 2050
2030 PRINT"WRONG-IT IS UNIT 5"
2040 GO TO 2060
2050 PRINT"CORRECT"
2060 END
```

*Example Problem - Program to calculate hydrogen ion concentrations*

Here is a simple program to calculate the approximate hydrogen ion concentration in an aqueous solution of a monoprotic acid

$$HA = H^+ + A^-$$

The equation to be solved is:
$$K = x^2/(c-x)$$
where x is the hydrogen ion concentration and c is the total acid concentration. Application of the quadratic formula gives us:

$$x = \frac{K \pm \sqrt{K^2 + 4Kc}}{2} \qquad \text{eqtn (1)}$$

If K is small (less than $10^{-6}$) and c is large, x can be further approximated by:

$$x = \sqrt{Kc} \qquad \text{eqtn (2)}$$

Bearing in mind that $\sqrt{\phantom{a}}$ is identical to raising to the power $\frac{1}{2}$ i.e. $a = a^{\frac{1}{2}}$, and that only the positive root of eqtn (1) is acceptable, we can plan a simple program:

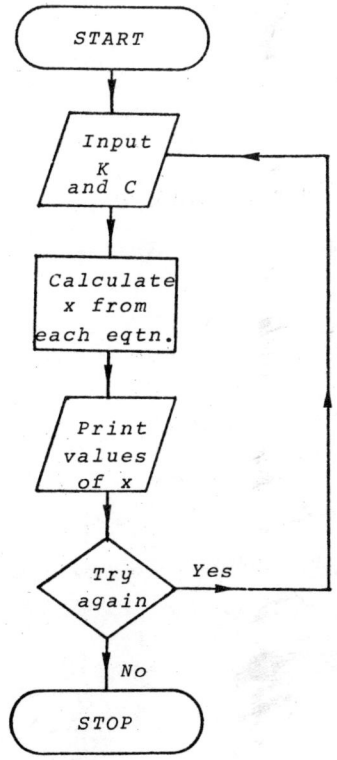

The detailed programming is overleaf.

```
100 PRINT"INPUT K";
110 INPUT K
120 PRINT"INPUT C";
130 INPUT C
140 LET A=(K↑2+4*K*C)↑0.5
150 LET X1=(-K+A)/2
160 LET X2=(K*C)↑0.5
170 PRINT"VALUE FROM EQUATION 1 =";X1;"    FROM EQUATION 2 =";X2
180 PRINT"TRY AGAIN - YES OR NO?";
190 INPUT S$
200 IF S$="YES" THEN 100
210 END

    RUN

INPUT K← 1.8E-5
INPUT C← 0.1
VALUE FROM EQUATION 1 = .133267E-2     FROM EQUATION 2 = .134164E-2
TRY AGAIN - YES OR NO?← YES
INPUT K← 1.8E-5
INPUT C← 0.001
VALUE FROM EQUATION 1 = .125466E-3     FROM EQUATION 2 = .134164E-3
TRY AGAIN - YES OR NO?← NO
```

Please ensure that you SAVE this program for the following practice problem.

---

*Practice Problem*

It is often useful to have built-in checks in a program so that, for example, carelessly typed input is rejected or queried. Bearing this in mind, modify the preceding program HPLUSCALC so that negative or zero values of $K$ and $c$ are not accepted.

*BASIC in Chemistry*                                                         31

## UNIT 6.   LOOPING

*Performance Objectives:   At the end of this unit you should be able to use:*
   *(1) simple loop techniques in your programs;*
   *(2) nested loops.*

---

Each program described so far has executed any particular series of statements <u>once only</u>.   We can however, execute a series as many times as we wish if we use the FOR....TO and NEXT couplet.   Compare the use of this complet, in the program sequence below, with the alternative version using an IF statement:

```
1020 FOR I=1 TO 10            100 LET I=1
1030 LET J=100/I              110 LET J=100/I
1040 PRINT J                  120 PRINT J
1050 NEXT I                   130 LET I=I+1
                              140 IF I<=100 THEN 110
```

Either of these alternatives would produce identical results. Therefore, in the version using FOR/NEXT, we find that the sequence of lines, 1030 and 1040, will be repeated or <u>looped</u> ten times.   The index I takes the values 1,2,3.... 10, and J would therefore, be printed as 100,50,33.3333 and so on. Line 1050 causes I to be incremented by unity and transfers control back to line 1020 until I exceeds 10, at which stage control is passed to the line following 1050.

The index of a loop can be incremented by numbers other than 1.   For example, if we replaced line 1020 by

$$\text{1020 FOR I=2 TO 10 STEP 2}$$

then I would take the values 2,4,6,8 and 10.   The STEP is assumed to be 1 unless specifically stated.

Neither the limits of the index nor the step need to be positive, so long as they are compatible with each other. For example, we could have:

$$\text{1100 FOR I = 5 TO -5 STEP -2}$$

but we could not have

$$\text{1105 FOR I = 5 to -5 STEP 2}$$

because this would attempt to increase I from 5.   Such an error would normally be reported to the user.

On a point of good programming practice, you should ensure that the limits of the index in a FOR statement are an exact multiple of the STEP size.   For example:

$$\text{1150 FOR J = 1 to 3 STEP 0.33}$$

Here   J would take the values 1, 1.33, 1.66, 1.99, 2.32, 2.65, 2.98 but would never be equal to 3!

When it is desired to simultaneously vary two or more indices in a sequence of statements, this is easily achieved with sets of <u>nested</u> loops. The index used in one loop must not be used in any loop in the same nest. Also inner loops must lie complete inside any outer ones: e.g.

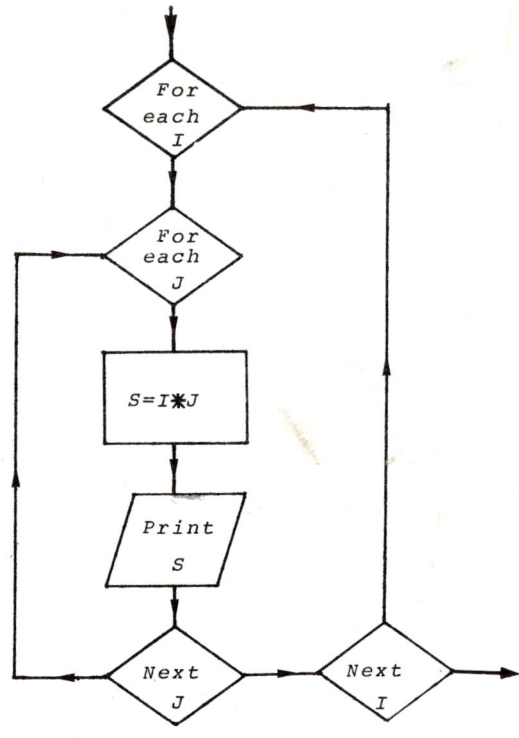

```
 90 FOR I=3 TO 5
100 FOR J=2 TO 6 STEP 2
110 LET S=I*J
120 PRINT S
130 NEXT J
140 NEXT I
```

This would print the values 6,12,18,8,16,24,10,20 and 30 because the inner loop runs over its entire index range for <u>each</u> value of the outer loop index.

These are examples of valid loop structures

```
┌→100 FOR I=1 TO 5
│   .....
│   ......
│ ┌→200 FOR J=1 TO 7
│ │   ......
│ └─800 NEXT J
│   ......
└──900 NEXT I
```

```
┌→100 FOR L=1 TO 4
│   .....
│ ┌→200 FOR M=1 TO 3
│ │   ......
│ │ ┌→300 FOR N=2 TO 9
│ │ │   ......
│ │ └─600 NEXT N
│ └──700 NEXT M
└────800 NEXT L
```

The following are not valid:

```
100 FOR I=1 TO 6              200 FOR I=1 TO 6
    ........                      ........
150 FOR I=1 TO 9              260 FOR J=1 TO 10
    ........                      ........
    ........                      ........
500 NEXT I                    700 NEXT I
                              800 NEXT J
```

---

*Example Problem - program to compute average values*

Presume that you have the results of n analyses for an element in a chemical compound.  The number of test data and their values are stored in the two DATA statements:

1090 DATA 6
1100 DATA 12.2,12.4,12.1,12.0,12.4,12.3

Here is a simple flowchart and program to calculate the average value of the six results (of course, as many results as desired could be used)

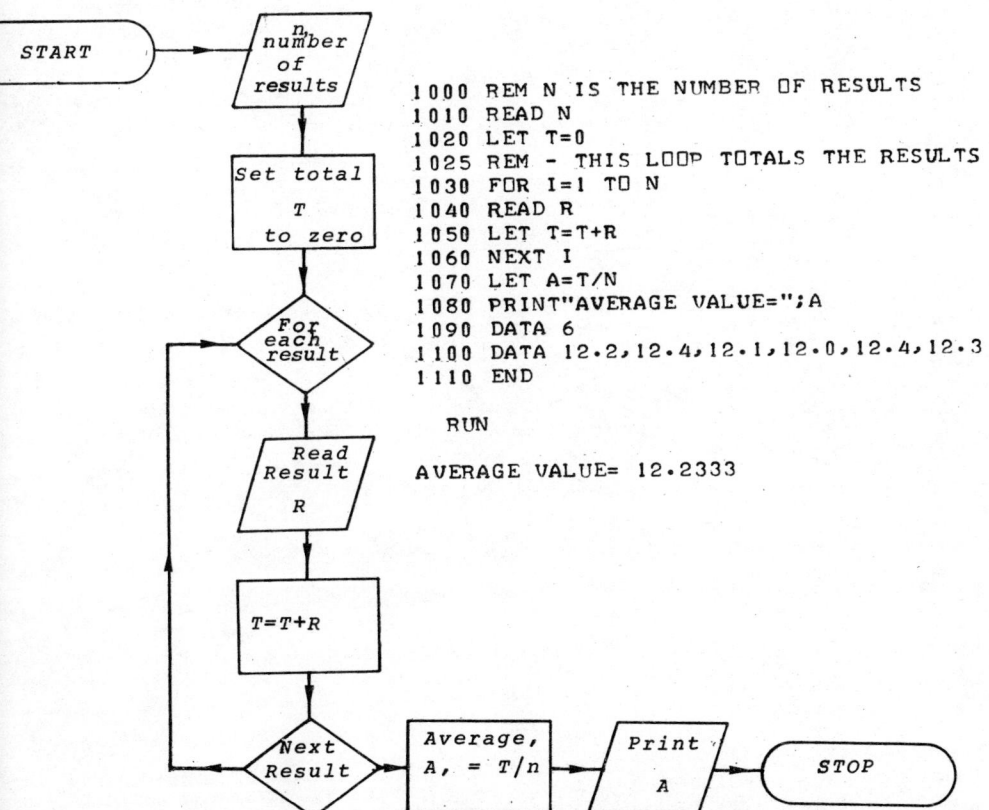

```
1000 REM N IS THE NUMBER OF RESULTS
1010 READ N
1020 LET T=0
1025 REM - THIS LOOP TOTALS THE RESULTS
1030 FOR I=1 TO N
1040 READ R
1050 LET T=T+R
1060 NEXT I
1070 LET A=T/N
1080 PRINT"AVERAGE VALUE=";A
1090 DATA 6
1100 DATA 12.2,12.4,12.1,12.0,12.4,12.3
1110 END

RUN

AVERAGE VALUE= 12.2333
```

*Practice problems*
(1) Using a similar method to that used in the previous program, write a program to compute individual averages for the analyses of three different batches of materials. For each analysis there will be a DATA statement to specify the number of repeat measurements for each batch, followed by a DATA statement containing the measurements:
```
1000 DATA 6
1010 DATA 12.2,12.4,12.1,12.0,12.4,12.2
1020 DATA 3
1030 DATA 28.1,28.0,28.3
1040 DATA 5
1050 DATA 14.4,14.5,14.3,14.6,14.3
```
Try to use just two nested FOR/NEXT loops to solve his problem.
(2) Use the program in the example section, and modify it so that both the mean (average) and standard deviation are calculated. Compute the latter from the formula
$$\sigma^2 = \Sigma_i (x_i - \mu)^2 / (n-1)$$
where $\sigma$ = standard deviation
$\mu$ = mean
$x_i$ = result of measurement i
$n$ = number of results
and $\Sigma$ means "sum over all of the values"

*BASIC in Chemistry* 35

## UNIT 7.   LISTS AND TABLES OF QUANTITIES

*Performance Objectives:   at the end of this unit you should be able to:*
  *(1) Store lists of variables;*
  *(2) Store tables of variables;*
  *(3) Manipulate these lists and tables and use them sequentially with FOR/NEXT loops.*

So far, each variable has been represented by a separate symbol such as A, C or X3.  This imposes great limitation on our programs which can be avoided by the use of subscripted variables similar to the familiar mathematical notation:

| mathematical | BASIC |
|---|---|
| $x_1, x_2, x_3 \ldots \ldots \ldots x_{20}$ | $X(1)\ X(2), X(3) \ldots \ldots X(20)$ |
| $a_{1,1}, a_{1,2}, \ldots a_{2,1}, a_{2,2} \ldots a_{4,5}$ | $A(1,1), A(1,2) \ldots A(2,1), A(2,2) \ldots A(4,5)$ |

The former is a list, the latter is a table.  In either case the quantities such as $x_1, x_2$ and $x_3$ are called "elements" so that $x_{10}$ is the tenth element in the list of x-values.  It is represented in BASIC by X(10); indeed, any item in the list can be represented by a variable such as X(I) and this is an example of a <u>singly-subscripted</u> variable - X being the variable and I the subscript.

The second example is a table with four rows and five columns:

$$\begin{array}{ccccc} a_{1,1} & a_{1,2} & a_{1,3} & a_{1,4} & a_{1,5} \\ a_{2,1} & a_{2,2} & a_{2,3} & a_{2,4} & a_{2,5} \\ a_{3,1} & a_{3,2} & a_{3,3} & a_{3,4} & a_{3,5} \\ a_{4,1} & a_{4,2} & a_{4,3} & a_{4,4} & a_{4,5} \end{array}$$

The element in, for example, the third row and fourth column is $a_{3,4}$.  This is represented by the <u>doubly subscripted</u> variable A(3,4); any chosen variable in the table could be referred to as A(I,J).

In either case the BASIC variable is comprised of a single letter followed by either one or two indices in parentheses. The indices can be written as integers or represented by integer variables such as I, J, D1 and so on.  For example, X(K), Y(P) and A(I,J) are all valid subscripted variables.  If the index is non-integer, it is truncated (e.g. from 4.9 to 4).

The maximum size of a list or table must be specified, usually at the beginning of a program, with a DIM statement.  Examples are 200 DIM X(20) - allows the list of variables X(0) to X(20) 345 DIM Y(50,5) - allows a table of 51 rows (0 to 50) and 6 (0 to 5) columns.  Dimensions of various lists and tables can be declared in a single DIM statement:
100 DIM X(20), Y(50,5), Z(9)

On a number of computers, the zero'th elements such as X(0), Y(0,0), Y(0,1) are not recognised so it is advisable not to try to use them.

Also, it is not usually essential to use DIM if the maximum subscript does not exceed 10. Many compilers permit dimensions of the form DIM X(N) where the value of N is specified later in the program. Whilst these are commonly available features, they are by no means universal and we therefore do not recommend their use.

---

*Example problem - program for least squares data fitting*

Experimental data consists of two types of quantities:
  those that we can measure - called dependent variables
  those that we can control - called independent variables

We will represent these variables by y and x respectively. An example could be the dependence of pH (y) of a solution on temperature (x).

Frequently, the data are described by a straight line passing through the experimental points as in the figure, below, which illustrates the dependence of y upon x. The equation of the

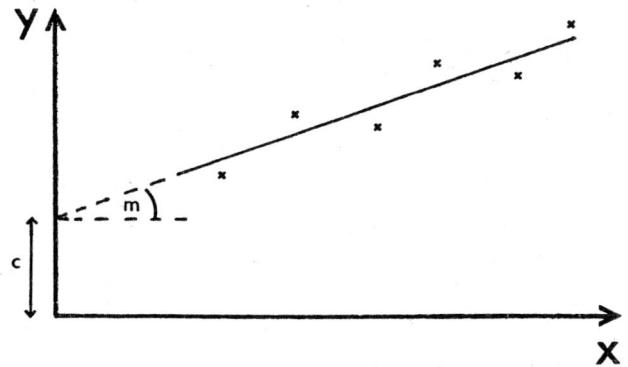

straight line is

$$y'_i = m.x'_i + c$$

where $x'_i$, $y'_i$ are coordinates of points on the straight line passing through the x, y data; m is the gradient ($\Delta y'/\Delta x'$) and c is the intercept on the y axis. The <u>best</u> straight line will be the one for which we have the smallest sum of the squares $(y'_i - y_i)^2$ for each of the data points. Some fairly simple mathematics based on this <u>least squares</u> premise gives us two simultaneous equations to solve:

$$\Sigma y_i = m \Sigma x_i + c.n$$
$$\Sigma x_i y_i = m \Sigma x_i^2 + c \Sigma x_i$$

where n is the number of data points and $\Sigma$ means "the sum of" - e.g. $\Sigma y_i$ is the sum of all the n experimental values of y. On solving the equations, we obtain:

$$m=(\Sigma y_i \Sigma x_i - n\Sigma x_i y_i)/\left[(\Sigma x_i)^2 - n\Sigma x_i^2\right]$$

$$c=(\Sigma x_i \Sigma x_i y_i - \Sigma y_i \Sigma x_i^2)/\left[(\Sigma x_i)^2 - n\Sigma x_i^2\right]$$

We will use these results to calculate the coordinates of the "best" straight line passing through our original data.

Here is a possible flowchart solution:

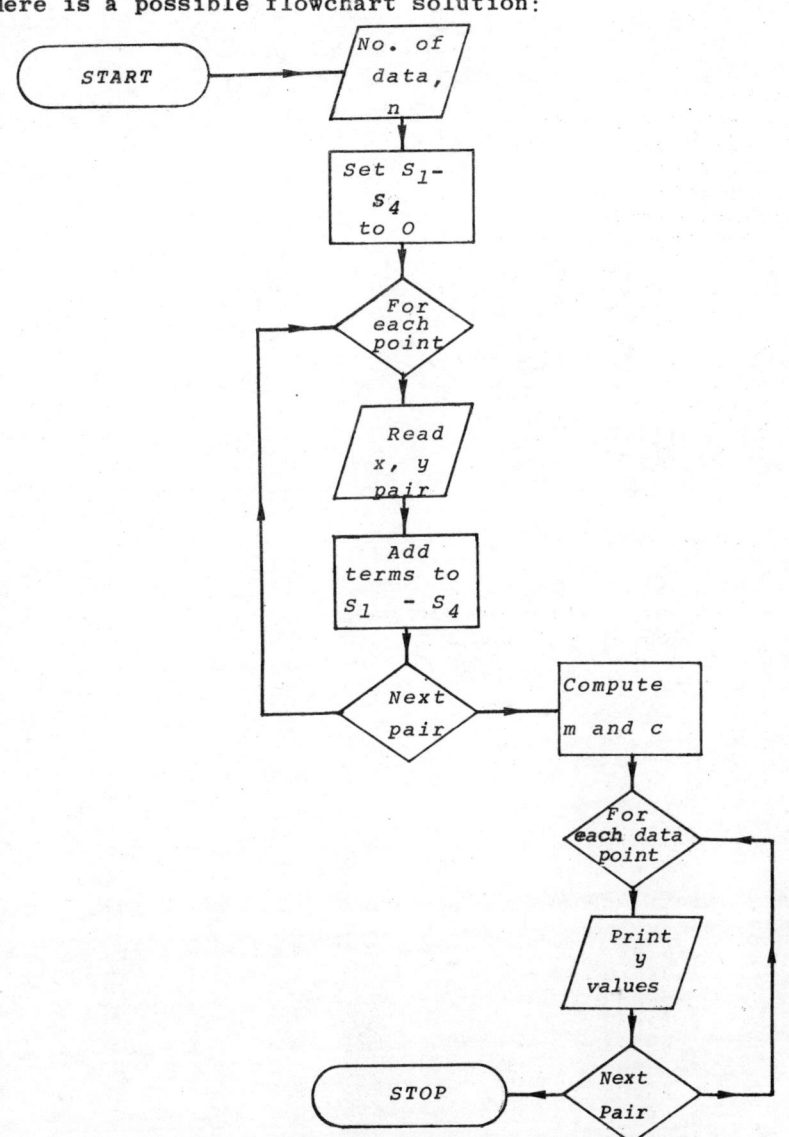

**A detailed program based on this is on the next page:**

```
1000 DIM X(100),Y(100),Z(100)
1005 PRINT"HOW MANY DATA POINTS ARE THERE?";
1010 INPUT N
1015 REM-INITIALISE SUMS TO ZERO
1020 LET S1=0
1030 LET S2=0
1040 LET S3=0
1042 LET S4=0
1045 REM-THIS LOOP SUMS THE VALUES OF X,Y,X*Y AND X↑2
1047 PRINT"INPUT DATA AS X,Y PAIRS-ONE PAIR PER LINE"
1048 PRINT"AND PUT A COMMA BETWEEN EACH VALUE"
1050 FOR I=1 TO N
1055 PRINT"INPUT X,Y PAIR";
1060 INPUT X(I),Y(I)
1070 LET S1=S1+Y(I)
1080 LET S2=S2+X(I)
1090 LET S3=S3+X(I)*Y(I)
1095 LET S4=S4+X(I)↑2
1100 NEXT I
1110 LET M=(S1*S2-N*S3)/(S2↑2-N*S4)
1120 LET C=(S2*S3-S1*S4)/(S2↑2-N*S4)
1130 PRINT
1135 PRINT"EXPERIMENTAL","LEAST SQUARES"
1140 PRINT"------------","----- -------"
1150 FOR I=1 TO N
1160 LET Z(I)=M*X(I)+C
1170 PRINT Y(I),Z(I)
1180 NEXT I
2000 END
```

A typical run is:

```
HOW MANY DATA POINTS ARE THERE?← 9
INPUT DATA AS X,Y PAIRS-ONE PAIR PER LINE
AND PUT A COMMA BETWEEN EACH VALUE
INPUT X,Y PAIR← 5.4,3.2
INPUT X,Y PAIR← 17.0,3.65
INPUT X,Y PAIR← 27.5,4.8
INPUT X,Y PAIR← 39.5,5.5
INPUT X,Y PAIR← 51.0,6.55
INPUT X,Y PAIR← 66.8,7.55
INPUT X,Y PAIR← 78.5,8.7
INPUT X,Y PAIR← 92.0,9.4
INPUT X,Y PAIR← 99.4,10.2

EXPERIMENTAL    LEAST SQUARES
------------    ----- -------
   3.2             3.0281
   3.65            3.90242
   4.8             4.69383
   5.5             5.5983
   6.55            6.46508
   7.55            7.65597
   8.7             8.53783
   9.4             9.55536
  10.2            10.1131
```

This program (LEASTSQ) illustrates a number of points:
>The use of DIM to allow up to 100 pairs of points
>the use of $S_1, S_2$ etc. as "accumulators" of intermediate results
>interaction between user and computer

This program will be modified in the next unit - so be sure to SAVE a copy of it.

---

*Practice Problems*

(1) Plot the experimental and least squares y values, from the preceding program, against the input x values and compare the least squares line with your estimate of the best straight line.

(2) Represent the summations in LEASTSQ by doubly-subscripted variables:

$$S(1,1) = S(2,2) = \Sigma x_i$$
$$S(1,2) = n$$
$$S(2,1) = \Sigma x_i^2$$

Modify the program, using the correct DIM statement and re-run it. (Although this example is artificial, it illustrates the use of double subscripts in tables).

## UNIT 8.   CHARACTER STORAGE AND MANIPULATION

*Performance objectives:   at the end of this unit you should be able to:*
   *(1) Manipulate string variables as if they were normal variables;*
   *(2) Use string variables to plot simple graphs on a TTY.*

Sequences of characters or <u>string variables</u> were first mentioned in UNIT 4.   Examples of valid string variable names are:
$$A\$  \quad S\$  \quad T\$(1)  \quad T\$(X)$$
However, names such as A1\$, \$B or \$ are all invalid.   Many compilers will store a string (usually no more than 15 characters) in a single variable.   So, for example, the variable C\$ could store the string SODIUM NITRATE (14 characters including a space) but not HYDROCHLORIC ACID (17 characters).   Other compilers require you to declare the maximum number of characters to be stored in a string variable so that, in our first example we would add a DIM statement at the start of our program
100 DIM C\$(14).
   This kind of dimensioning is acceptable on just about all machines, though you should be careful not to confuse this with the use of <u>string array elements</u> which may each be able to store up to 15 characters on some compilers.
   String variables can be used in many circumstances:

```
100 LET A$="YES"              A$ contains the string YES
500 READ B$,C$                Assigns YES and NO to B$ and
999 DATA "YES","NO"           C$
250 INPUT B$                  Whatever is typed, e.g. YES is
                              assigned to B$
450 PRINT B$,C$               Prints the contents of B$ and
                              C$, e.g.
                              YES                     NO
220 IF A$="YES" THEN 600      Transfers control to line 60
                              if A$ does contain YES
320 IF A$<B$ THEN 500         The strings are compared in an
                              alphabetical dictionary sense.
                              If A$ contains FIRST and B$
                              contains FOREMOST control is
                              transferred to line 500
```

   The safest relational tests are for equality and inequality but the "dictionary" comparison is widely available.   This comparison is made in precisely the way that we would refer to words in a conventional English dictionary.
   Many compilers permit "sub-string" manipulations which enable users to access part of a character string.   Some other compilers permit both string arrays <u>and</u> sub-strings - check on the availability of these features on your machine.
   Another use of characters is in plotting simple graphs on a TTY.   These are not very accurate but are useful for a quick visual display as can be seen from the following program.

*Example problem - program to plot a simple graph*

We will plot, on the TTY, a simple graph to show the dependence of the experimental and least squares values of y on x obtained from UNIT 7. It is easiest to plot the graph "sideways on" so that the printer head position is proportional to y and the amount of paper fed through the TTY is proportional to x. If we decide to use 50 horizontal print positions to represent y, then we simply scale our y values to lie between 1 and 50. Then, on each line we might print * and O at the correct relative positions to represent experimental and computed values of y.

The following flowchart illustrates the modifications to be made to program LEASTSQ, based on the preceding discussion:

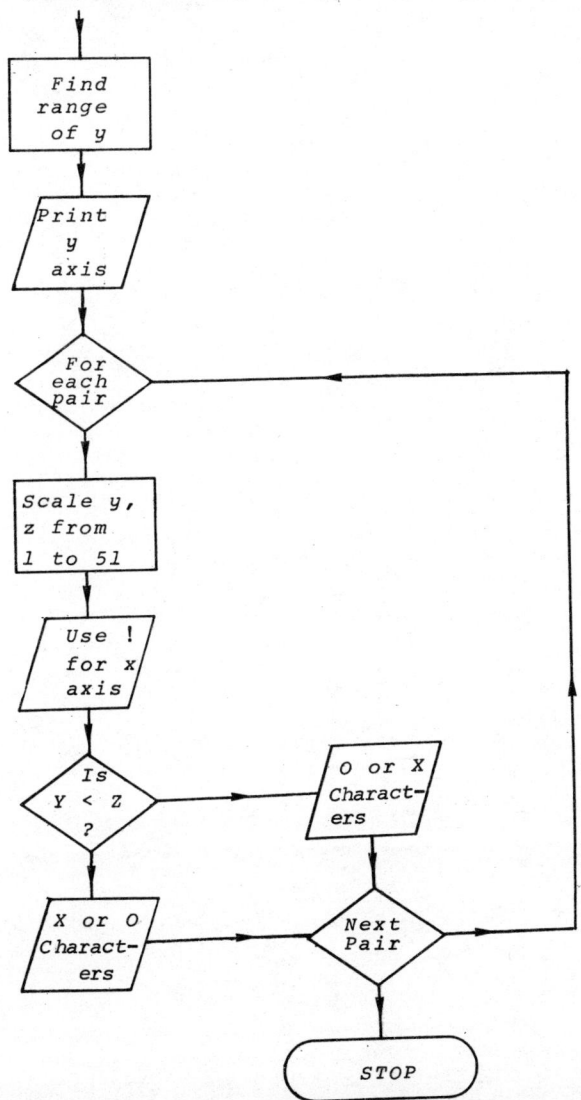

The corresponding lines of BASIC coding that need to be inserted between lines 1180 and 2000 of LEASTSQ are:

```
1190 LET L=10↑6
1200 LET U=-10↑6
1210 FOR I=1 TO N
1220 IF Y(I)>=L THEN1240
1230 LET L=Y(I)
1240 IF Z(I)>=L THEN 1260
1250 LET L=Z(I)
1260 IF Y(I)<=U THEN 1280
1270 LET U=Y(I)
1280 IF Z(I)<=U THEN 1300
1290 LET U=Z(I)
1300 NEXT I
1305 REM-PRINT THE Y AXIS HORIZONTALLY
1308 PRINT"GRAPH WILL BE PLOTTED WITH O=EXPERIMENTAL AND X= LEAST"
1309 PRINT"SQUARES.ONLY X OR O PRINTED AT ANY ONE POSITION"
1310 PRINT
1312 PRINT"L-------------------------------------------------------U"
1315 REM-CALCULATE THE SCALE FACTOR
1320 LET S=50/(U-L)
1330 FOR I =1 TO N
1335 REM-CALCULATE Y COORDS BETWEEN 1 AND 51 INCLUSIVE
1340 LET P1=(Y(I)-L)*S+1
1350 LET P2=(Z(I)-L)*S+1
1355 LET E=INT(P1)
1360 LET F=INT(P2)
1365 LET A$="O"
1370 LET B$="X"
1380 PRINT"!"
1390 PRINT"!"
1400 PRINT"!";
1410 IF P1<=P2 THEN 1460
1420 LET E=INT(P2)
1430 LET F=INT(P1)
1440 LET A$="X"
1450 LET B$="O"
1460 FOR J=1 TO 51
1470 IF J<>E THEN 1500
1480 REM-PRINT STATEMENT TAKES ACCOUNT OF X-AXIS
1490 PRINT TAB(J+1);A$;
1500 IF J<>F THEN 1530
1510 PRINT TAB(J+1);B$;
1520 GO TO 1540
1530 PRINT TAB(J+1);" ";
1540 NEXT J
1560 PRINT
1570 NEXT I
1580 PRINT
1590 PRINT"GRAPH PLOTTED OVER RANGE OF Y FROM L=";L;"TO U=";U
2000 END
```

*BASIC in Chemistry*

Note that INT takes the integer part of the number which follows in brackets - lines 1500 and 1510, (see UNIT 9). It is necessary to use INT so that integer values can be stored in E and F. If we insert these lines into the original program and use the same x,y data then we will, of course, obtain exactly the same numerical output as before. Also, we get the simple graphical output which is reproduced below:

```
GRAPH WILL BE PLOTTED WITH O=EXPERIMENTAL AND X= LEAST
SQUARES.ONLY X OR O PRINTED AT ANY ONE POSITION

L--------------------------------------------------U
!
!
!  X O
!
!
!       O   X
!
!
!             X  O
!
!
!                OX
!
!
!                    X  O
!
!
!                         O X
!
!
!                             X O
!
!
!                                  O X
!
!
!                                        X O
!
GRAPH PLOTTED OVER RANGE OF Y FROM L= 3.0281   TO U= 10.2
```

Clearly, the graph is very approximate because the TTY resolution is plus or minus one character position. It is, however, particularly useful as a quick check prior to using a more accurate plotting device.

---

*Practice Problems*
(1) Ensure that you can successfully modify LEASTSQ to give graphical output, as described above.
(2) The radioactivity of a sample was measured (in counts per minute) 37 times and the results were found to be distributed in the following ranges:

| Range: | 1440-1450 | 1450-1460 | 1460-1470 | 1470-1480 | 1480-1490 |
|---|---|---|---|---|---|
| No. within range | 3 | 8 | 15 | 7 | 4 |

Write a short program to represent this distribution in the form of a simple histogram, in which the lengths of the horizontal bars are proportional to the number of results. For example:

1440***  
1450********  
1460***************  
1470*******  
1480****  
1490

## UNIT 9.  BASIC SHORTHAND

*Performance Objectives:*  At the end of this unit, you should be able to:
  (1) Use standard BASIC functions such as SQR to obtain square roots;
  (2) Define functions to carry out specific mathematical processes.

Many mathematical procedures such as calculations of square roots and taking logarithms of numbers can be accomplished in a single BASIC statement.  The conversion of numbers into integers was an example of this technique used in the preceding unit.  Some other examples are:

```
1010 LET A=SQR(B)        assigns the square root of B to A
1020 LET Y=LOG(A)        assigns the natural (base e) logarithm
                         of A to Y
```

These <u>function statements</u> are of the general form

$$\text{function name (argument)}$$

The function name describes the mathematical function to be carried out on the argument which can be a numerical quantity or an arithmetical expression.  Some further examples of common functions are:

| Function | Example | Meaning |
|---|---|---|
| SQR(x) | 50 LET S=SQR(16) | Square root of 16 assigned to S. |
| INT(x) | 65 LET Q=INT(2.9)<br>70 LET Q=INT(-4.1) | Assigns value of argument rounded down to the nearest whole number (Q=2 and -5 in the examples). |
| ABS(x) | 20 LET T=ABS(3-8) | Assigns absolute value (i.e. negative sign ignored).  T becomes +5 in the example. |
| LOG(x) | 30 LET B1=LOG(3.2) | B1 assigned value of log 3.2. |
| EXP(x) | 140 LET C=EXP(4.8) | Sets C to $e^{4.8}$ where e=2.718. |
| RND(0) | 149 LET E=RND(0) | Selects random numbers in the range 0 to 1 inclusive. |
| RND(-1) | 17 LET F=RND(-1) | If argument is 0 the same series of numbers is generated for each run; if it is -1, the sequence is different. |
| SIN(x)<br>COS(x)<br>TAN(x) | 78 LET L=SIN(0.4)<br>86 LET L=COS(A)<br>800 LET L=TAN(1-C/3) | Assigns the sine, cosine or tangent of the angle x in radians to L. |

With the exception of the RND function, an arithmetical argument can take any form. For example:

340 LET C=SQR(T↑2 +L*Q/T1+ 3.28)

Functions can also be used in place of variables:

270 LET R=(LOG(387)+SQR(T))/INT(T+0.5)

You can also <u>define</u> a function for some common operation that may be unique to your program. For this you use the DEF statement. This must be of the form

DEF FNx(y)=expression

x and y must be single letters. The second letter is called a "dummy variable" and can also appear in the expression on the right which should not exceed one line. The defined function must not appear in the expression - that is, the function can not be used in a recursive manner. An example of a defined function is

500 DEF FNS(Y)=Y/(4+Y)

Later we could give Y a value:

1000 LET A=FNS(2)

This assigns the value of 2/(4 +2) to A. Clearly, Y, the dummy variable is substituted by whatever is enclosed in brackets when we use our function FNS( ).

---

*Example Problem - Computer simulation of experimental data*

The generation of simulated experimental data is a worth while application. Provided that we have a suitable theoretical equation, parameters for the chosen experiment can be put into a program and, from these, simulated data can be generated with, if desired, any degree of simulated random scatter. Furthermore, each set of data can be generated under hypothetical conditions that may be impossible to obtain in the laboratory, such as very high pressures or temperatures. This is, of course, only true if the theoretical equation is still applicable under the chosen experimental conditions.

As an example, let us generate vapour pressure data at various absolute temperatures. A suitable equation* is

$$\log_{10} p = b - 0.05223a/T \quad \text{or} \quad p = 10^{(b-0.05223a/T)}$$

where p is the vapour pressure (in mm of mercury), T is the absolute temperature and a and b are constants for any chosen liquid. A flowchart to generate p values with a ±5% random scatter is shown on the next page.

---

*Handbook of Chemistry and Physics (published by The Chemical Rubber Co.) 49th Edition, p. D139.

## BASIC in Chemistry

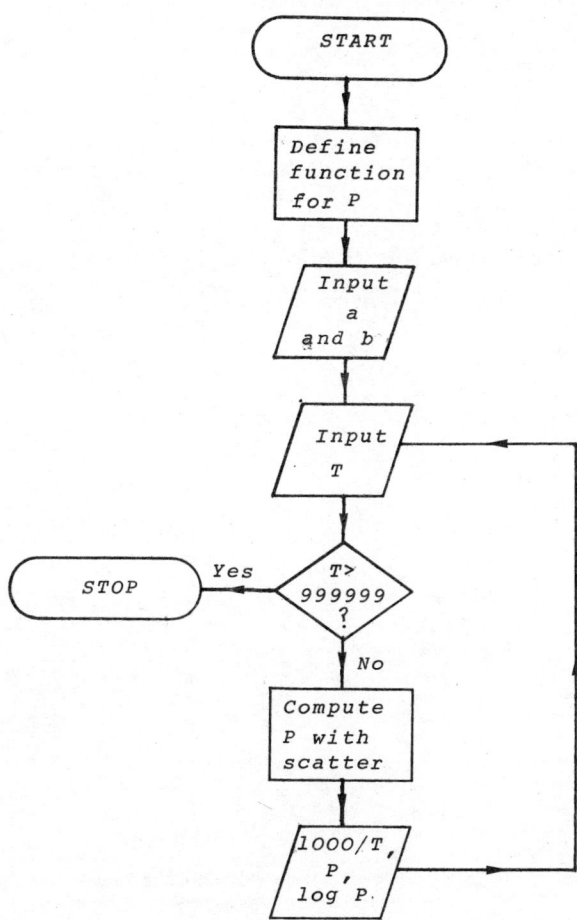

A program based on this flowchart is shown below:

```
70 REM-DEFINE THE FUNCTION TO BE USED
80 DEF FNP(T)=10↑(B-0.05223*A/T)
90 PRINT"TO STOP THE PROGRAM INPUT 999999 FOR THE TEMPERATURE"
95 PRINT"INPUT A AND B VALUES";
97 INPUT A,B
100 PRINT"INPUT TEMPERATURE";
110 INPUT T
115 IF T=999999 THEN 500
120 LET T1=1000/T
130 LET P=FNP(T)
135 REM-ADD BETWEEN + AND - 5% RANDOMLY TO P
140 LET D=(RND(-1)-0.5)/10
150 LET P=P+D*P
160 LET L=LOG(P)
170 PRINT"1000/T =";T1;" PRESSURE/MM HG =";P;"LOG P=";L
180 GO TO 100
500 END
```

Some output from this program is shown below:

```
TO STOP THE PROGRAM INPUT 999999 FOR THE TEMPERATURE
INPUT A AND B VALUES- 33914,8.004
INPUT TEMPERATURE- 293
1000/T = 3.41297   PRESSURE/MM HG = 88.6406 LOG P= 4.48459
INPUT TEMPERATURE- 303
1000/T = 3.30033   PRESSURE/MM HG = 141.571 LOG P= 4.9528
INPUT TEMPERATURE- 313
1000/T = 3.19489   PRESSURE/MM HG = 228.679 LOG P= 5.43232
INPUT TEMPERATURE- 323
1000/T = 3.09598   PRESSURE/MM HG = 319.494 LOG P= 5.76674
INPUT TEMPERATURE- 333
1000/T = 3.003     PRESSURE/MM HG = 490.163 LOG P= 6.19474
INPUT TEMPERATURE- 999999
```

The values of a and b are for carbon tetrachloride. Other values for a variety of liquids are listed in the "Handbook of Chemistry & Physics". The program can be modified by the use of a different function for the vapour pressure and comparisons can be made with experimental data to select the most appropriate equation.

*Practice Problems*

(1) Take some other laboratory experiment and write a program to simulate the results that would be obtained, using a similar approach to that used in the program above (examples are: radioactive decay; acid-base titration; pH titrations; a simple NMR spectrum with a TTY plot).

(2) Write a program to calculate and, if possible, to plot a simple TTY graph of the orbital:

$$\psi_{1s} = \frac{1}{\sqrt{\pi}} \left(\frac{Z}{a_o}\right)^{3/2} e^{-Zr/a_o}$$

(Z is the nuclear charge; $a_o$ is 0.529 Angstroms and r is the distance from the nucleus in Angstroms).

Define a function using DEF for the 1s orbital and use this in the program. It is then a simple matter to substitute other orbitals. Since the 1s orbital has spherical symmetry, a graph can be plotted with the nucleus at the centre of the picture, and r as the only variable. Non-spherically symmetrical orbitals such as $p_x$ are a little more difficult because of their angular dependence.

## UNIT 10. HOW TO RE-USE STATEMENTS

*Performance Objectives: at the end of this unit you should be able to:*
*(1) Decide when subroutines should be used instead of defined functions;*
*(2) Write and use subroutines in your programs.*

If a routine calculation can not be written in one line with a DEF function, then the GOSUB/RETURN couplet may be the solution. This is similar to the use of the GO TO statement but GOSUB only temporarily interrupts the program sequence, transfers control to a section of the program called <u>subroutine</u>, and then returns to the line following GOSUB. The subroutine may be entered many times by a program. Furthermore, one subroutine may call another although this feature should be used with care.

An example is:

```
      480 GOSUB 630----,
  ,->490 PRINT Q       |
  |    ......          |
  |  630 LET A=B/C <---'
  |    ....                  ] Subroutine
  |    .....
  |  690 LET Q=A+B+C
  '---700 RETURN
```

The line number referred to in the GOSUB statement corresponds to the first line in the sequence of lines in the subroutine which must be terminated by RETURN. In this case, control is then transferred to line 490. Values of all variables needed in the subroutine must, of course, be assigned before the GOSUB.

Subroutines can be used more generally in a technique called "structured programming". This is beyond the scope of this book, but details may be found in the article by S. J. Carpenter and K. R. Knight in the "International Journal of Mathematical Education in Science and Technology" Vol. 5, No.3, p539 (1974).

*Example Problem: Temperature corrections to standard enthalpies*

The heat capacity, $C_p$, of a substance can be represented by an equation such as

$$C_p = a + bT + cT^{-2}$$

(Alternatively, a polynomial such as $e + fT + gT^2$ can be used). This can be used to calculate enthalpies and enthalpy changes at temperatures other than the standard one of 298K. For one substance, its enthalpy at a temperature T is given by

$$H_T = H_{298} + \int_{298}^{T} (a + bT + cT^{-2})\, dT$$

or $H_T = H_{298} + a(T-298) + b(T^2 - 298^2)/2 - c(1/T - 1/298)$

Therefore, for any reaction, we can calculate the corrected enthalpies for each reactant and product and, by subtraction, calculate the net enthalpy of reaction. This is a useful application of the subroutine concept since the same calculations are carried out for reactants and products in different parts of the program. A possible flowchart is shown below:

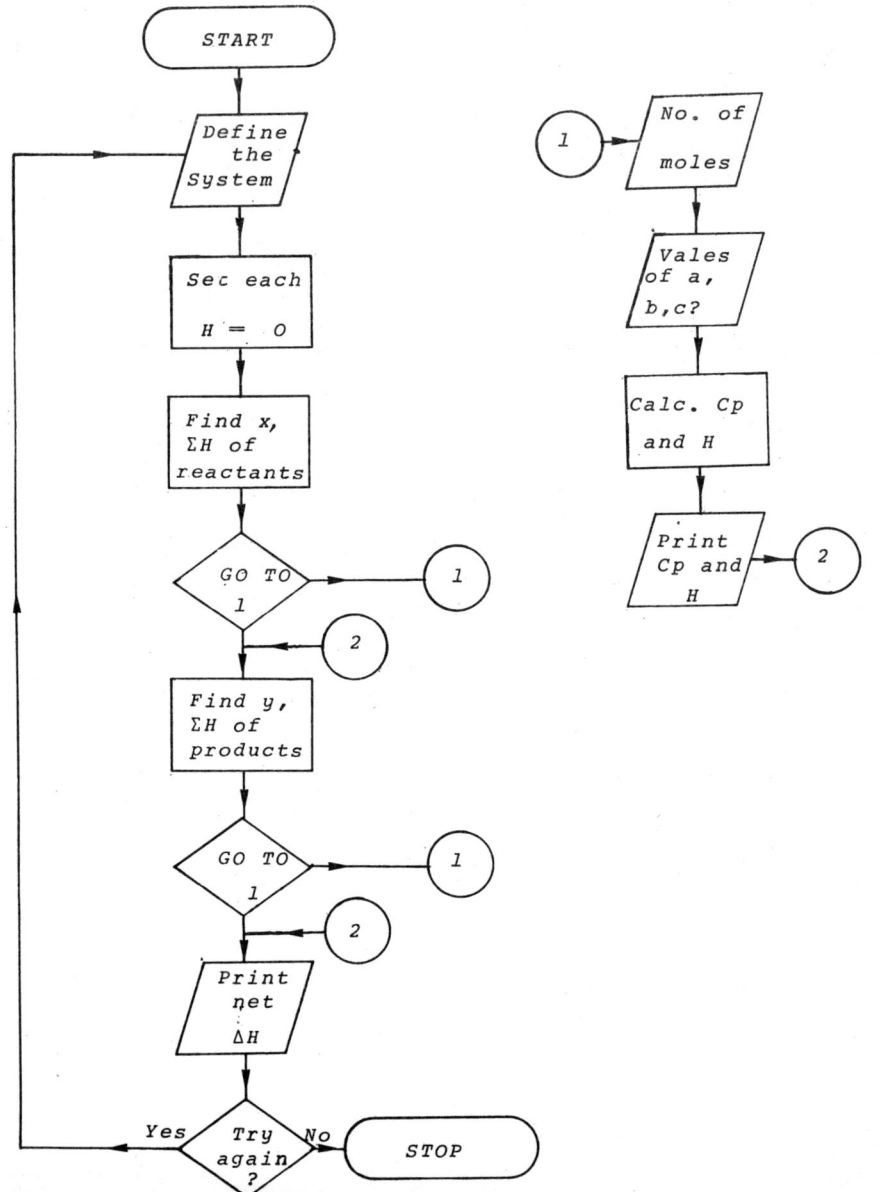

The corresponding BASIC program is as follows: (some compilers will require A$ to be dimensioned with DIM A$(3))

```
100 PRINT"HOW MANY REACTANTS?";
110 INPUT M
120 PRINT"HOW MANY PRODUCTS?";
130 INPUT N
140 PRINT"ABSOLUTE TEMPERATURE?";
150 INPUT T
160 PRINT"ASSUME      CP=A+B*T+C*T↑(-2)"
170 LET X=0
180 LET Y=0
190 FOR I=1 TO M
200 PRINT"STANDARD ENTHALPY OF REACTANT";I;"IN KJ MOL-1?";
210 GO SUB 400
220 REM-SUBROUTINE RETURNS VALUES OF V AND H AT TEMPERATURE T
230 LET X=X+H2*V
240 PRINT
250 NEXT I
260 FOR J=1 TO N
270 PRINT"STANDARD ENTHALPY OF PRODUCT";J;"IN KJ MOL-1?";
280 GO SUB 400
290 LET Y=Y+H2*V
300 PRINT
310 NEXT J
320 LET H=Y-X
330 PRINT
340 PRINT
350 PRINT"ENTHALPY OF REACTION AT";T;"DEGREES IS";H;"KJ MOL-1"
360 PRINT"TRY AGAIN-YES OR NO?";
370 INPUT A$
380 IF A$="YES" THEN 100
390 GO TO 530
    400 REM-HERE IS THE SUBROUTINE
    410 INPUT H1
    420 PRINT"HOW MANY MOLES";
    430 INPUT V
    440 PRINT"VALUES OF A,B AND C?";
    450 INPUT A,B,C
    460 LET C1=A+B*T+C*T↑(-2)
    470 LET Q=A*(T-298)+(B/2)*((T↑2)-298↑2)-C*((1/T)-1/298)
    480 LET H2=H1+Q/1000
    490 PRINT"VALUE OF CP IS ";C1;" J K-1 MOL-1"
    500 PRINT"ENTHALPY IS ";H2;" KJ MOL-1"
    510 REM-NOW RETURN TO THE MAIN PROGRAM
    520 RETURN
530 END
```

Notice that it is conventional to indent the subroutine section so that it can be easily identified.
An example of the use of this program is provided overleaf, for the reaction:

$$SO_2(g) + \tfrac{1}{2}O_2(g) \rightarrow SO_3(g)$$

```
RUN

HOW MANY REACTANTS?← 2
HOW MANY PRODUCTS?← 1
ABSOLUTE TEMPERATURE?← 673
ASSUME     CP=A+B*T+C*T↑(-2)
STANDARD ENTHALPY OF REACTANT 1 IN KJ MOL-1?← -296.85
HOW MANY MOLES← 1
VALUES OF A,B AND C?← 43.429,10.63E-5,-5.941E5
VALUE OF CP IS   42.1889   J K-1 MOL-1
ENTHALPY IS -281.656   KJ MOL-1

STANDARD ENTHALPY OF REACTANT 2 IN KJ MOL-1?← 0
HOW MANY MOLES← 0.5
VALUES OF A,B AND C?← 29.957,4.184E-3,-1.673E5
VALUE OF CP IS   32.4035   J K-1 MOL-1
ENTHALPY IS   11.6828   KJ MOL-1

STANDARD ENTHALPY OF PRODUCT 1 IN KJ MOL-1?← -394.97
HOW MANY MOLES← 1
VALUES OF A,B AND C?← 57.32,26.86E-3,-13.05E5
VALUE OF CP IS   72.5155   J K-1 MOL-1
ENTHALPY IS -371.025   KJ MOL-1


ENTHALPY OF REACTION AT 673   DEGREES IS-95.2107 KJ MOL-1
TRY AGAIN-YES OR NO?← NO
```

---

*Practice problems*

(1) The number of combinations from m items taken n at a time is represented by $\binom{m}{n}$ and can be calculated from:

$$\binom{m}{n} = \frac{m!}{n!\,(m-n)!}$$

where, for example, $m!$ equals $m.(m-1).(m-2).\,\text{---}\,.1$
Write a program to calculate $\binom{m}{n}$ and use a subroutine to calculate $m!, n!$ and $(m-n)!$

(2) Most computer systems permit a subroutine for one program to be used in another. Find out if this can be done on your computer and write a subroutine to solve two simultaneous equations. This will be useful in many areas of physical chemistry and spectroscopy – for example in multicomponent analysis.

## UNIT 11. ELEGANCE AND SOPHISTICATION

*Performance Objectives; at the end of this unit you should be able to:*
  *(1) Use doubly-subscripted variables;*
  *(2) Use matrix operations and statements where they are more efficient than other BASIC statements.*

In UNIT 7 we introduced the idea of subscripted variables. Any list of variables can be represented by single subscripts such as A(1), A(2) A(3) etc.  Often, however, data is arranged in a table consisting of rows and columns such as:

|     |   | \column 1 | 2 | 3 | 4 |
|-----|---|------|-----|-----|-----|
|     | 1 | 87.4 | 0.7 | 6.3 | 5.2 |
| row | 2 | 6.2  | 9.0 | 0.9 | 2.2 |
|     | 3 | 55.9 | 0.6 | 4.4 | 2.9 |
|     | 4 | 5.6  | 5.9 | 2.3 | 8.4 |

This has been briefly discussed in UNIT 7 and you will know therefore, that any item in the table can be selected by its row and column numbers.  Hence, a doubly subscripted array of variables can be used to store such a table of values.  We could use variables B(I,J) so that, for example, the variable B(2,3) would be 0.9 in our example since it is conventional to specify the row number first and the column number last (I and J respectively).  A table of values of the type described is called a <u>matrix</u>; a matrix with equal numbers of rows and columns is said to be <u>square</u>.  A list of singly-subscripted variables should be called a <u>vector</u> but for our purposes we will regard it as a matrix with just one column.

Most BASIC compilers let you carry out some very elegant programming with matrices and the associated <u>matrix operations</u>. Any BASIC statement which carries out a matrix operation contains the word MAT as, for example, in MAT READ which allows you to read in the entire contents of a matrix (list or table) in a single statement.

To illustrate this, presume that a program must read all of the values of four subscripted variables.  Compare the two ways of doing this:

```
     Using FOR and NEXT            Using MAT READ
   120 DIM A(4)
   130 FOR I=1 TO 4             120 DIM A(4)
   140 READ A(I)                130 MAT READ A
   150 NEXT I                   900 DATA 2.8,3.2,6,8.9
   900 DATA 2.8,3.2,6,8.9
```

Notice that, because of the automatic nature of the MAT statements, you have to be particularly careful in the use of DIM statements.  If, in the above example we had

120 DIM A(100)

then the statement in line 130 would attempt to read 100 items of data.  This can be circumvented by writing, in this case,

130 MAT READ A(4)

You should, however, avoid any redimensioning of this type since, not only will you then need to keep careful track of each change of dimension, but also this trick is not allowed with some BASIC compilers. Remember also, that all matrix statements ignore elements with zero subscripts - that is, variables as A(0) A(0,1) and so on, do not exist so far as we are concerned.

Continuing with our examples, we can print the values of our list, A, with a single MAT PRINT statement, rather than a FOR/NEXT loop:

                        220 MAT PRINT A

This will print each value of A(I) on a new line, since the statement prints the contents of a matrix a row at a time, and our list is a matrix with just one column and each row contains just one item of data.

Similarly, the contents of a table can be read or printed:

        *Using FOR/NEXT*            *Using MAT READ*

```
1000 DIM B(3,4)
1010 FOR I=1 TO 3
1020 FOR J=1 TO 4                 1000 DIM B(3,4)
1030 READ B(I,J)                  1010 MAT READ B
1040 NEXT J
1050 NEXT I
```

In each case, DATA statements are needed and the contents of the matrix will be entered row-by-row. Similarly, a MAT PRINT will print the contents of B row-by-row. In this case,

```
2000 MAT PRINT B      would print each element 15 spaces apart
2000 MAT PRINT B;     would leave one to three spaces, depending
                      on the computer used (see UNIT 4)
```

Matrices can be added, subtracted or multiplied in a single MAT statement. Compare the two ways of adding together the corresponding elements of two tables:

      *Using FOR/NEXT*        *Using MAT*

```
500 FOR I=1 TO N
510 FOR J=1 TO N                  500 MAT S= A+B
520 LET S(I,J)=A(I,J) + B(I,J)
530 NEXT J
540 NEXT I
```

Similarly we could write

              345 MAT D = A - B

In each case the matrices must have the same dimensions (same number of rows, same number of columns) which have previously been declared in a DIM statement.

Matrix multiplication is achieved by statements such as:

           150 MAT R=S*T           250 MAT A =B*B

The same matrix must not be present on each side of the equals sign. Matrices are multiplied in the mathematical sense; for

example, if matrix B is of dimensions (1,3) i.e. a single
row, and if C has dimensions (3,1) i.e. a single column of
singly subscripted variables, then we have:

$$\text{MAT } P = B*C = \begin{bmatrix} b_{11} & b_{12} & b_{13} \end{bmatrix} \begin{bmatrix} c_{11} \\ c_{21} \\ c_{31} \end{bmatrix} = b_{11}c_{11} + b_{12}c_{21} + b_{13}c_{31}$$

So that matrix P has dimensions (1,1), just one row and one
column, respectively.
If, on the other hand, the dimensions are (2,3) and (3,2)
respectively then the matrix multiplication is as follows:

$$\text{MAT } P = B*C =$$

$$\begin{bmatrix} b_{11} & b_{12} & b_{13} \\ b_{21} & b_{22} & b_{23} \end{bmatrix} \begin{bmatrix} c_{11} & c_{12} \\ c_{21} & c_{22} \\ c_{31} & c_{32} \end{bmatrix} = \begin{bmatrix} b_{11}c_{11}+b_{12}c_{21}+b_{13}c_{31} & b_{11}c_{12}+b_{12}c_{22}+b_{13}c_{32} \\ b_{21}c_{11}+b_{22}c_{21}+b_{23}c_{31} & b_{21}c_{12}+b_{22}c_{22}+b_{23}c_{32} \end{bmatrix}$$

Matrix P now has dimensions (2,2) and so, by implication, this
requires the dimensions of the matrices being multiplied to
<u>conform</u>. In general, the product matrix P will have dimensions
(m,n) if the first matrix B has dimensions (m,i) and the second
has dimensions (i,n). Check that this is true for the above
examples.

Each element of a matrix can be multiplied by a constant,
which must be enclosed in brackets:

    150 MAT A=(2.5)*A          560 MAT Q= (X↑2 + 5)*C

In this case, the same matrix can appear on both sides of the
equals sign.

A matrix can be initialized (all elements set to zero) with
the ZER statement:

    200 MAT Y = ZER

All elements are set to unity by CON:

    400 MAT X = CON

A useful statement is IDN, which can only be used on a square
matrix. This sets each diagonal element to unity and all
others to zero; this is an example of such a matrix:

$$A = \begin{bmatrix} 1 & 0 & 0 & 0 & 0 & 0 & 0 \\ 0 & 1 & 0 & 0 & 0 & 0 & 0 \\ 0 & 0 & 1 & 0 & 0 & 0 & 0 \\ 0 & 0 & 0 & 1 & 0 & 0 & 0 \\ 0 & 0 & 0 & 0 & 1 & 0 & 0 \\ 0 & 0 & 0 & 0 & 0 & 1 & 0 \\ 0 & 0 & 0 & 0 & 0 & 0 & 1 \end{bmatrix}$$

To generate this we would write

    200 MAT A =IDN

The concept of an identity matrix is of some importance, as will be shown shortly.
One of the most powerful features of matrix statements in BASIC is the ease with which <u>matrix inversion</u> can be programmed. If $Q$ is a square matrix, then the inverse of $Q$ (denoted by $Q^{-1}$) is a matrix which, when multiplied by $Q$ yields an identity matrix:

$$Q \cdot Q^{-1} = \begin{bmatrix} 1 & 0 & \cdots & & \cdots & 0 \\ 0 & 1 & \cdots & & \cdots & 0 \\ \cdots & \cdots & & & \cdots & \\ \cdots & \cdots & & & \cdots & \\ 0 & \cdots & & \cdots & 1 & 0 \\ 0 & \cdots & & \cdots & 0 & 1 \end{bmatrix}$$

The laborious mathematics of matrix inversion are eliminated by the INV statement, for example:

230 MAT R= INV(Q)

The matrix $R$ must have the same dimensions as $Q$ and now contains the inverse of $Q$. Note that the same matrix must not appear on both sides of the equals sign.

Finally, one can transpose a matrix such that each element $a_{i,j}$ becomes $b_{j,i}$. For example, G is the transpose of F:

$$F = \begin{bmatrix} 2 & 8 & 1 \\ 9 & 3 & -4 \\ 5 & 7 & 6 \end{bmatrix} \qquad G = \begin{bmatrix} 2 & 9 & 5 \\ 8 & 3 & 7 \\ 1 & -4 & 6 \end{bmatrix}$$

For this, we use the TRN statement:

180 MAT G = TRN(F)

Some of these MAT statements will be used in the following problem.

---

*Example Problem—using a quadratic equation for least squares fitting*

Experimental data ($y_i$) are often fitted to a quadratic equation in $x_i$, the dependent variable at data point i. The fitted value of $y_i$ is $f_i$:

$$f_i = a_0 + a_1 x_i + a_2 x_i^2$$

Given the values of x and y, what are the values of $a_0, a_1$ and $a_2$? If we knew these, we could calculate the "best" values of the fitted variables, $f_i$, that pass through our data points, as shown overleaf:

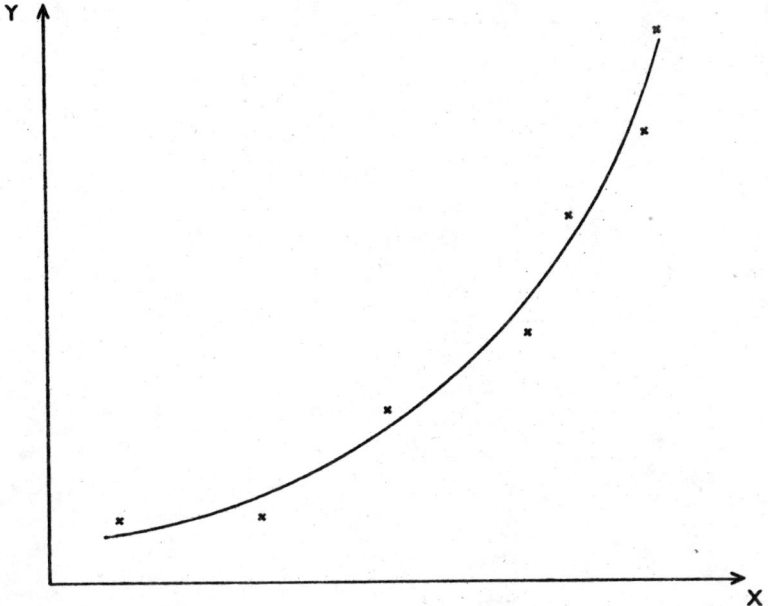

The "least-squares" method (UNIT 7) can be used to solve this problem and, after a simple mathematical treatment, a set of three simultaneous equations in the three unknowns $a_0, a_1$ and $a_2$ is obtained:

$$\Sigma y_i = a_0 N + a_1 \Sigma x_i + a_2 \Sigma x_i^2$$

$$\Sigma x_i y_i = a_0 \Sigma x_i + a_1 \Sigma x_i^2 + a_2 \Sigma x_i^3$$

$$\Sigma x_i^2 y_i = a_0 \Sigma x_i^2 + a_1 \Sigma x_i^3 + a_2 \Sigma x_i^4$$

The equations can be abbreviated to

$$p_1 = a_0 q_{11} + a_1 q_{12} + a_2 q_{13}$$

$$p_2 = a_0 q_{21} + a_1 q_{22} + a_2 q_{23}$$

$$p_3 = a_0 q_{31} + a_1 q_{32} + a_2 q_{33}$$

In matrix notation, we have

$$P = Q \cdot A$$

By using the inverse of $Q$, we can find the matrix $A$ (which contains the values of $a_0$, $a_1$ and $a_2$):

$$(Q^{-1}).P = (Q^{-1}.Q).A$$

$$= A$$

A flowchart based on the above is shown below:

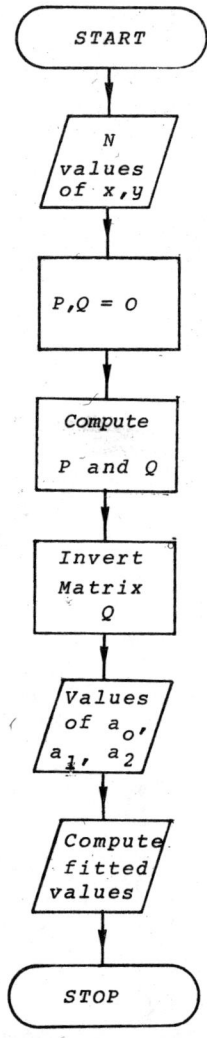

A program based on this flowchart is shown on the next page. Note that we make use of the symmetry of the $Q$ matrix to avoid duplication of our calculations. Also, remember that matrix operations ignore the zero'th elements so that the parameters $a_0$, $a_1$ and $a_2$ are actually stored in the elements $A(1)$, $A(2)$ and $A(3)$. Finally, notice that the three DATA statements contain the number of points, the list of x values and the list of y values, respectively.

```
100 DIM X(100),Y(100)
105 DIM A(3),P(3),Q(3,3),T(3,3)
120 READ N
130 MAT READ X(N)
140 MAT READ Y(N)
145 REM-SET ALL MATRIX ELEMENTS TO ZERO
150 MAT P=ZER
160 MAT Q=ZER
165 REM - TOTAL ALL THE SUMS
170 FOR I =1 TO N
180 LET P(1)=P(1)+Y(I)
190 LET P(2)=P(2)+X(I)*Y(I)
200 LET P(3)=P(3)+Y(I)*X(I)↑2
210 LET Q(1,2)=Q(1,2)+X(I)
220 LET Q(1,3)=Q(1,3)+X(I)↑2
230 LET Q(2,3)=Q(2,3)+X(I)↑3
240 LET Q(3,3)=Q(3,3)+X(I)↑4
250 NEXT I
255 REM-USE THE SYMMETRY OF THE EQTNS TO CALCULATE THE OTHER SUMS
260 LET Q(1,1)=N
270 LET Q(2,1)=Q(1,2)
280 LET Q(3,1)=Q(1,3)
290 LET Q(3,2)=Q(2,3)
300 LET Q(2,2)=Q(1,3)
305 REM-INVERT MATRIX Q
310 MAT T =INV(Q)
320 MAT A=T*P
330 PRINT"VALUES OF A ARE";
340 MAT PRINT A
350 PRINT"EXPERIMENTAL","LEAST SQUARES"
360 PRINT"-------------","--------------"
370 FOR I= 1 TO N
380 LET Y1=A(1)+A(2)*X(I)+A(3)*X(I)↑2
390 PRINT Y(I),Y1
400 NEXT I
405 DATA 8
410 DATA 20,30,40,50,55,60,65,70
420 DATA 1.7,2.1,2.8,3.85,4.5,5.25,6.9,8.0
500 END
```

The data contained in the DATA statements (lines 405, 410 and 420) gave the following results when the program was run:

```
RUN

VALUES OF A ARE

 3.23466
-.1248
 .273734E-2

EXPERIMENTAL    LEAST SQUARES
------------    -------------
    1.7             1.8336
    2.1             1.95427
    2.8             2.6224
    3.85            3.83801
    4.5             4.65111
    5.25            5.60108
    6.9             6.68791
    8               7.91162
```

*Practice Problems*

(1) Plot out the results from the above program either manually or on the TTY (see UNIT 8) to show the goodness of fit to the experimental data.

(2) A quadratic fitting routine of the type described needs at least three data points, and preferably more if the answers are to be meaningful. Build in a check to prevent less than three data being input and print a warning message.

(3) We often need to solve two or more simultaneous equations (see practice problem 2, UNIT 10). Use MAT statements and write a subroutine that will solve up to 10 simultaneous equations and test it in a small program with the following equations:

$$x_1 + x_2 + x_3 + x_4 + x_5 = 15$$

$$x_1 - x_2 - x_3 + x_4 + x_5 = 5$$

$$2x_1 + x_2 - x_3 + 2x_4 - x_5 = 4$$

$$x_1 + x_2 - 2x_3 - 3x_4 + x_5 = -10$$

$$x_1 + 2x_2 + 3x_3 + 4x_4 - 2x_5 = 20$$

## UNIT 12.  FILING USEFUL DATA

*Performance Objectives:* at the end of this unit you should be able to:
    (1) Store numerical data in a BASIC file;
    (2) Access the data for processing.

    Data for a BASIC program can be stored in a <u>file</u> on many computers, so that large amounts of data can be input without the somewhat laborious usage of simple READ or INPUT.  This permits, for example the collection of data on paper tape from an instrument for later analysis.  At the outset, it must be emphasized that these facilities are by no means universal and the detailed procedures show considerable variation between different machines.  Data can be stored in files in a separate part of the computer memory in sequential order - exactly as it is required in a program.
    A file of data is created in one of two ways:
    (1) from a BASIC program by printing the data into a specified file;
    (2) by using the computer operating system for storing any type of file.  This is particularly useful with large amounts of experimental data which may either be obtained "online" for immediate storage in a file or stored on some computer readable medium such as paper tape.

In either case, the data file is identical.  There are, however, some slight variations in the precise manner in which the data is stored for a BASIC program on different computers.  For example, consider this simple program:

```
100 READ A,B,C,D
110 READ X,Y,Z
120 LET Q = (A+B)/C+D)
130 LET R = X*Y*Z
140 PRINT Q,R
150 DATA 180,200,198,32
160 DATA 3.2,4.8,1.9
170 END
```

The values in the DATA statements can be stored in a file which might be called FILEA but this might have one of the two forms:

    150 180,200,198,32
    160 3.2,4.8,1.9    or    180,200,198,32,3.2,4.8,1.9

The two differ only in the presence or absence of line numbers.  The latter is the more common and, in addition, the commas are often not required.  For convenience, we shall emphasize this type of file construction and usage.  You should check on the local conventions for file usage on your computer.
    If your computer is able to use files for BASIC programs, then you must specify the file required in a special declaration statement.  For example, we were able to use the FILES statement.  This is used with the special file-reading statements as shown in the following lines, which replace lines 100 and 110 in the program above:

    90 FILES FILEA    (tells the computer that FILEA is needed)

    100 READ #1,A,B,C,D
    110 READ #1,X,Y,Z

A file reading statement of this type is of the form:

    READ #m, list

If there is only one file of data to be accessed by the program, as in our example, then m is equal to one. If two or more files are being used, m is equal to an integer corresponding to the order in which a file was delcared. For example, the data in FILEA could be divided so that:

    FILEB could contain the data   180 200 198 32
    and FILEC could contain         3.2 4.8 1.9

To allow for this our program would need the FILES statement:

    90 FILES FILEB, FILEC

We would access the data by file read statements of the form:

    100 READ #1,A,B,C,D   (reads data from FILEB)
    110 READ #2,X,Y,Z     (reads data from FILEC)

Thus, m is an index which specifies the file to be used.
    If data is being read continuously from a file, we will need to know when the end of the file has been reached which will, at least, prevent us from attempting to read beyond the end of our list! For example, this couplet will keep trying endlessly to read a value for the variable A:

    600 READ #1,A
    .....
    700 GO TO 600

We could add, at the end of our file of data, a "check" value such as 999999 and, by testing for this value, the reading sequence could be terminated. It is easier, however, to use the IF END statement:

    600 READ #1,A
    610 IF END #1 THEN 710
    .......
    700 GO TO 600
    710 .....

The IF END #m statement transfers control to the specified line number after the last item of data has been read from file m.
    Data can only be read from a file in a serial manner - one item after another. If we want to read it all again from the beginning we use a RESTORE statement after the first READ and before the second. In the above example, we could write

    900 RESTORE #1

and follow this with further READ statements - compare this with RESTORE in UNIT 3.
    A file is normally in the "READ mode" - which means that data can be read from it but it is protected from being over written. We may wish to store computed results in a file for later use, so how can this be achieved? The answer is to use the SCRATCH #m statement which allows us to transcribe information using the WRITE #m statement. For example, the following program would enable us to overwrite the original contents of FILEB with

new values of X, Y and Z:

```
    90 FILES FILEA, FILEB
   100 READ #1,A,B,C,D
   110 READ #2,X,Y,Z
       .......                      New values of X, Y and Z
       ........                     are calculated
   180 SCRATCH #2                   FILEB changed to "WRITE mode"
   190 WRITE #2,X,Y,Z               Store new values in FILEB
```

If the new contents of FILEB are to be used in the same program, the file must be restored to the "READ mode" with the RESTORE command:

```
       ........
   500 RESTORE #2
   510 READ #2,X,Y,Z
       .....
```

Files are particularly useful when large amounts of data are to be processed; bearing this in mind, we will look at the illustrative program in our example problem section.

---
*Example Problem - peak location in chromatography*

Many instruments, such as gas chromatographs, produce a voltage output which varies with time. The problems are to:

  find accurate peak positions;
  discriminate between "noise" and genuine peaks;
  calculate, possibly after a curve resolving routine,
      the peak areas;
  correct for baseline shift or curvature.

We will use artificial, noise free, data with a level baseline to illustrate just the first of these applications. Also we shall presume that analogue - to digital conversion has already been carried out so that a series of voltage readings at equal time intervals is now stored in the computer in a file called NUFILE. The artificial chromatogram looked like this:

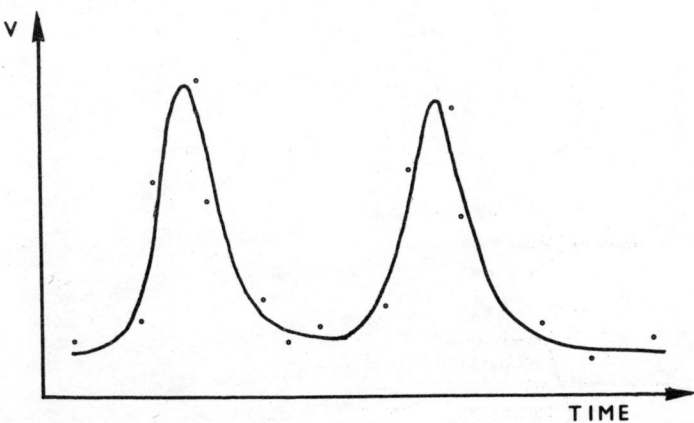

After conversion to digital form, the voltage readings at one second intervals were:

.1 .2 .7 1.4 2.4 4.0 7.5 13.8 11.2 6.8 3.9 2.3 1.1 .4 .2 .1 .2
0.0 .1 .3 .9 2.1 4.6 9.3 15.1 10.9 3.3 .3

This is now the data sequence in NUFILE.
We will now construct a program to find the peak locations.
So that small "jiggles" are not misinterpreted we will only identify a peak when the voltage has passed through a maximum <u>and</u> when it exceeds a pre-set threshold value. A suitable flowchart is shown below, with the corresponding program overleaf.

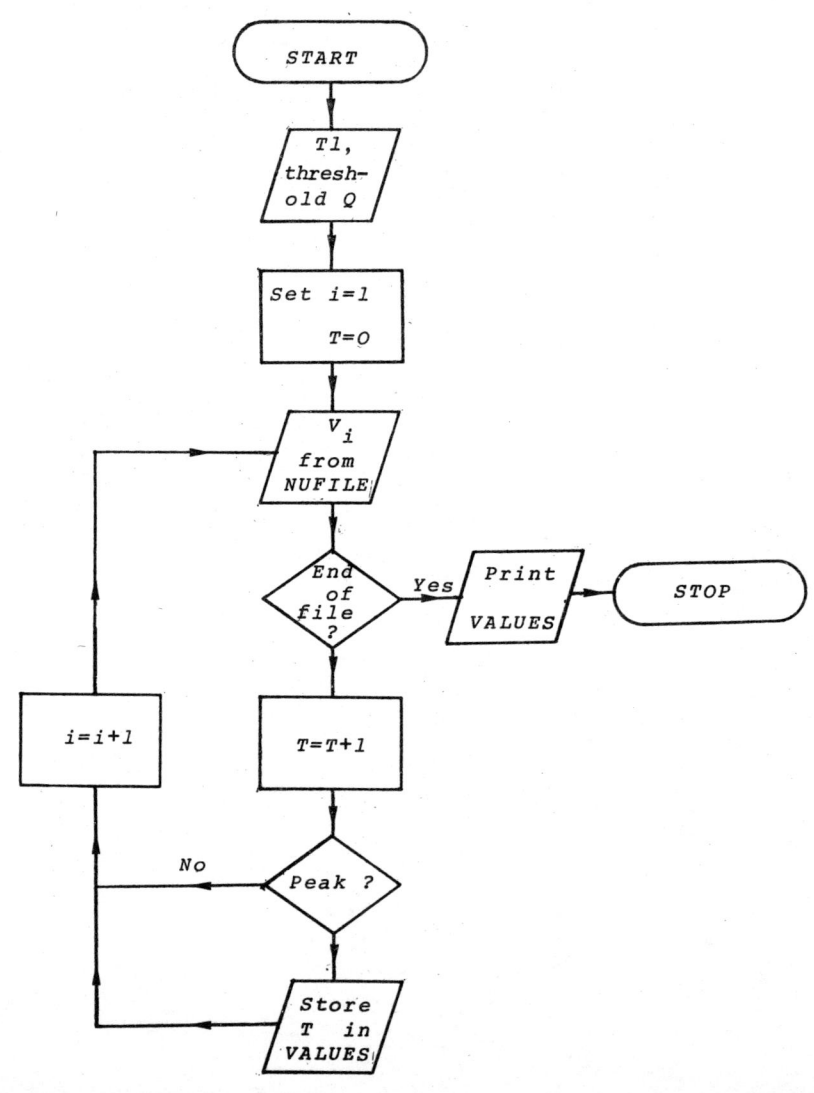

```
100 DIM V(3)
110 FILES NUFILE,VALUES
115 REM-SET FILE #2 TO THE READ MODE
120 SCRATCH #2
125 PRINT"INPUT THE THRESHOLD VALUE";
127 INPUT Q
130 PRINT"INPUT TIMING INTERVAL";
140 INPUT T1
150 LET N=0
160 LET T=T1
170 FOR J=1 TO 3
180 READ #1,V(J)
185 IF END#1 THEN 900
190 NEXT J
200 T=T+T1
205 REM-LINES 210 TO 230 DETECT PRESENCE OF A PEAK
210 IF V(2)<Q THEN 240
220 IF V(1)>=V(2) THEN 240
230 IF V(3)>=V(2) THEN 240
235 WRITE #2,T
240 LET V(1)=V(2)
242 LET V(2)=V(3)
245 READ #1,V(3)
247 IF END#1 THEN 270
250 LET N=N+1
260 GO TO 200
265 REM-NOW SET FILE#2 TO THE READ MODE
270 RESTORE#2
280 PRINT"THESE ARE THE T VALUES OF THE PEAKS"
290 FOR I= 1 TO N
300 READ #2,T
305 IF END#2 THEN 900
310 PRINT T;
320 NEXT I
900 END

              RUN PROCEEDING
INPUT THE THRESHOLD VALUE- 0.5
INPUT TIMING INTERVAL- 1
THESE ARE THE T VALUES OF THE PEAKS
 8   25
```

In this example, the data was idealized and, therefore, the program was very simple. Real-life data needs more sophisticated methods of analysis and these are mentioned in Part 3.

*Practice Problems*

(1) Find out how to create the file NUFILE in program above. Then ensure that you can use the file to obtain the same results that were quoted above.
(2) Modify the program so that it also calculates peak areas.
(3) If you are able to collect data from an analytical instrument (e.g. on paper tape) read the references in section 3.1.7 and write a program that will reduce instrumental noise and calculate the change in "signal to noise" ratio (reference 80). Also, devise a procedure for peak area apportionment.

## CONCLUSION TO PART 2: PROGRAMMING GUIDELINES

The following suggestions are additional to the BASIC conventions introduced in Units 1 to 12 and are intended as guidelines for the writing of more efficient and useful programs:

*General*

Use plenty of REM statements to explain, for example, the meanings of certain symbols or some tricky computation.
Do not unnecessarily evaluate an expression several times in different places - evaluate it once and store it as a variable.
Use the E notation for powers of 10 rather than writing, e.g. $6.3*10\uparrow4$. This has to be evaluated so write 6.3E4 instead.
Remember that the time taken by the compiler for a given operation increases in the order: addition (or subtraction); multiplication; division; exponentiation. So for example, A+A is evaluated more quickly than 2*A; 10*A more quickly than A/0.1 and A*A more quickly than $A\uparrow2$.
If an array element is to be used several times (e.g. in a loop) store it as a constant (e.g. A(2,4) could be stored as X)
BASIC only uses "real" arithmetic and does not use or permit integers. Hence, after some detailed arithmetic, if A is <u>computed</u> from an expression and B is some pre-set integer value, you might wish to test for the equality of A and B. For example,

   300 LET A= $(A\uparrow2)/(A*A)$
   400 IF A=B THEN 100

If B is equal to 1, then the test for equality will <u>always fail</u> because, due to truncation errors, A will be slightly different from unity. The problem can be partially circumvented by the use of INT (UNITS 8 and 9).

*Don't Depend on the Compiler to Bail You Out*

When a variable is being used as an accumulator or is being formed as a sum, be sure to set it to zero (if that is its intended initial value) at the beginning of the program.
Check input data to see that it falls between reasonable bounds. Similarly, check relevant computed values so that such errors as division by zero or taking logarithms of negative numbers are detected before they cause the program to fail.
Use subscript values for all array elements, even if not strictly necessary e.g. A(1) not just A.

*Write in a Modular Fashion*

Put all non-executable statements (e.g. DIM, DATA, FILES) into groups - preferably at the beginning of the program.
Ensure that common sections of coding are in subroutines and that these are indented for easy identification. Re-use subroutines in other programs when possible.

# PART THREE
## Applications

---

**3.1 A Survey of the Chemical Computing Literature**

Now that you have mastered the fundamentals of BASIC,, you can tackle many more advanced problems. To help you in selecting or solving suitable problems, we present, in this section, a survey of programming applications in seven selected areas. In each case, references are cited for the interested reader so that he can see how others have programmed their solutions to particular problems. Although FORTRAN and other languages are quoted in the references, BASIC can often be substituted with advantage.

*3.1.1 Concepts and General Strategy*

BASIC interactive packages find considerable utility in providing 'drill and practice' routines. This is a remedial use of the computer which, typically, presents repetitive examples of one type of problem to the student. The problems may be identical with just a random variation of data, thus only requiring different numerical answers. The program may continue in this manner or, at a more sophisticated level, it may present progressively more difficult problems based on the performance of the student. In more flexible systems, the program will give hints and suggestions to assist the student. Examples of this type are given by Breneman[1]. A typical dialogue could be (student responses are underlined):

```
WHAT IS THE BASE 10 LOGARITHM OF 625?      5.25
WRONG - NEED HELP - YES OR NO?             YES
O.K. - FIND THE LOG OF 6.25?               0.796
GOOD.  YOUR NUMBER IS 625 WHICH IS
100 TIMES LARGER THAN 6.25 SO YOU ADD THE
LOG OF 100 WHICH IS 2.   NOW TRY AGAIN.
WHAT IS THE BASE 10 LOGARITHM OF 625?      2.796
GOOD
 . . . . .
 . . . . .
   . . . .
  . . .
```

Such diagnostic conversations use subroutines (UNIT 10) in which only the input and output data differ. The technique is only slightly more difficult to apply to non-numeric problems.

Some applications are:
- Mole-gram conversion[1]
- Ideal gas calculations[1,2]
- Qualitative inorganic analysis[3]
- Balancing chemical equations[4,5]
- Organic nomenclature[6]
- Organic syntheses[6,7]
- Organic stereochemistry[6,8]
- Electronic configurations[9]

## 3.1.2 Chemical Equilibria and Electrochemistry

This heading covers a number of areas, but electrochemistry has achieved some popularity. For example, changes in pH or d(pH)/d(volume) can be calculated and equivalence points located[10], together with successive ionisation constants. It is also instructive to compare the values of $H^+$ concentrations calculated from various approximations[11]. Strategies can be developed to enable students to explore simulated equilibrium systems and the papers by Runquist[12] and Craig[13] and their coworkers are of particular interest. More specific references are:

- Potentiometric titrations[14]
- Acid-base titration curve calculations[14,15,16]
- pH of weak acids, bases or their salts[17]
- pK values from spectroscopy[18]
- Stability constant for the $FeSCN^{2+}$ complex[19]
- Lanthanide-induced NMR shifts[20]
- Debye-Hückel theory[21]
- Ion selective electrodes[22,23]
- Gran's method for equivalence points[24]

## 3.1.3 Kinetics

Many programs have been published for the analysis of kinetic data by least-squares fitting techniques; the article by Williams and Taylor[25] is representative. The recent emphasis on simulation in kinetics is more relevant for our purposes because it enables the student to investigate the rates of chemical systems under conditions which may not be attainable in the laboratory. Examples of straightforward numerical simulations are quoted by Child[26] for the reaction between $U^{3+}$ and $[CrBr(H_2O)_5]^+$ and by Cabrol et al[27] for the reaction between acrolein and butadiene.

The Monte Carlo simulation technique is a powerful tool, particularly in view of its statistical nature which is superficially similar to our accepted picture of the way in which chemical reactions proceed. Examples in the literature include:

Simulation of complex schemes[28] such as

$$A \rightleftharpoons B \rightleftharpoons C \rightleftharpoons D$$

- Radioactive decay[29]
- Fundamental theory of gases, liquids and solids[30]
- Simulations with graph-plotter output[31]

A fairly up-to-date bibliography on the uses of computers in chemical kinetics has been written by Hogg[32] and this contains many more references than those presented here.

### 3.1.4 Quantum Theory

The numerical aspects of this subject are somewhat more difficult than those mentioned in the previous section. To reinforce the pure numerical aspects it has, therefore, become common practice to display features such as electron densities by means of simple graphs on peripheral devices such as printers, graph plotters or, if speed is not critical, on the TTY (see problem 2 page 43). Some examples of this and the many other possible applications include:

Rutherford scattering[33]
Photoelectric effect[34]
Contour mapping of orbitals[35,36]
Calculation of atomic energy levels[37]
Solutions to the Schrödinger equation[38,39]
Hartree-Fock methods[40]
Diatomic molecular orbital (m.o.) theory[41]
Huckel m.o. theory[42,43,44]

In connection with the last application, there is an instance of on-line computing as a lecture demonstration in the teaching of Huckel theory[45].

### 3.1.5 Spectroscopy

Computers are frequently used in the analysis of spectra, some examples are:
Infra-red spectra of diatomic molecules[46,47]
Identification of mass spectrum peaks[48]
Analysis of gamma ray spectra[49]
Atomic absorption spectrophotometry[50]

More recently, simulation and comparison techniques have been used and some examples in this area include:

Simulation of absorption spectra[51]
Interpretation of mass spectra[52,53,54]
Simulation of NMR spectra[55,56]
Plotting of NMR spectra[57,58]
Simulation of electron spin resonance spectra[59]

### 3.1.6 Statistics and Curve Fitting

The extreme numeracy of these subjects can be complemented, by computer assistance, with an emphasis on presentation and applications. For example, several aspects of curve fitting applications are described by Johnson[60], including Newton-Rapheson, Gauss-Seidel and Runge-Kutta methods. Other data reduction methods which have been implemented on computers are described in detail by Brevington[61].

Visual presentation of statistical properties of data (e.g. by histograms) is a good use of teletype facilities[62] and many other techniques have been used in statistics-related courses[63,64]. Some specific examples include:

Teletype representations of statistics[65]
Linear least squares fitting[66,67,68]
Non linear least squares fitting[69]
Statistical design of experiments[70]

Our list in this section would not be complete if we did not mention the users of computers for student grading of chemistry courses (and others!). For example:

Homework grading[71]
Grading of quantitative analysis reports[72,73]
Test and examination grading[74,75,76,77]
Questionnaire tabulation[78]

### 3.1.7 Analysis of Digitized Experimental Data

Most teletypes permit the reading of data from paper tape. Since instrumental data is fairly easily converted into digital form, suitable for paper tape output, numerous experiments are possible which illustrate the strategies of digital data treatment. An example has already been quoted in UNIT 12. Whilst digitisation is applicable to a wide range of experiments, there are many techniques common to most situations. The article by Perone and Eagleston[79], although mainly concerned with gas chromatography, discusses a number of general points. The article by Savitzky and Golay[80] is of particular importance as it discusses methods of data smoothing and peak location in considerable detail. Some other relevant examples include applications in:

Gamma-ray spectrometry[81]
Liquid scintillation counting[82]
Nuclear magnetic resonance[83]
Case history of laboratory automation[84]
Recent advances in data acquisition devices[85]

Note that we have disregarded mass spectrometry because of the somewhat expensive equipment needed for data acquisition[86].

## REFERENCES

1) G. L. Breneman, *J. Chem. Educ.*, **50**, 473 (1973)

2) D. F. Dever, *J. Chem. Educ.*, **51**, 650 (1974)

3) L. D. Francis, *J. Chem.*, **50**, 556 (1973)

4) R. S. Ratney, *J. Chem. Educ.*, **47**, 136 (1970)

5) R. C. Grandey, *J. Chem. Educ.*, **48**, 791 (1971)

6) L. B. Rodewald, G. H. Culp and J. J. Lagowski, *J. Chem. Educ.*, **47**, 134 (1970)

7) H. A. Clark, J. C. Marshall and T. L. Isenhour, *J. Chem. Educ.*, **50**, 645 (1973)

8) K. B. Wiberg, *J. Chem. Educ.*, **47**, 113 (1970)

9) G. L. Breneman, *J. Chem. Educ.*, **52**, 295 (1975)

10) R. E. Jensen, R. G. Garvey and B. A. Paulson, *J. Chem. Educ.*, **47**, 147 (1970)

11) J. E. House and R. C. Reiter, *J. Chem. Educ.*, **45**, 679 (1968)

12) O. Runquist, R. Olsen and B. Snadden, *J. Chem. Educ.*, **49**, 265 (1972)

13) N. C. Craig, D. D. Sheretz, T. S. Carlton and M. N. Ackerman, *J. Chem. Educ.*, **48**, 311 (1971)

14) A. R. Emery, *J. Chem. Educ.*, **42**, 131 (1965)

15) H. R. Ellison, *J. Chem. Educ.*, **51**, 738 (1974)

16) C. V. Bishop and M. A. Wartell, *J. Chem. Educ.*, **52**, 187 (1975)

17) G. G. Schlessinger, *J. Chem. Educ.*, **46**, 680 (1969)

18) R. J. Palma and J. W. Meux, *J. Chem. Educ.*, **51**, 448 (1974)

19) A. J. McCall Jr., *J. Chem. Educ.*, **52**, 118 (1975)

20) A. J. Rafalski and J. Barciszewski, *J. Mol. Struct.*, **19**, 223 (1973)

21) A. A. Zimmerman, *J. Chem. Educ.*, **47**, 146 (1969)

22) A. F. Isbell and R. L. Pecsok, *Anal. Chem.*, **45**, 2363 (1973)

23) M. J. D. Branch and G. A. Rechnitz, Anal. Chem., **42**, 1172 (1970)

24) T. J. MacDonald, B. J. Barker and J. A. Caruso, J. Chem. Educ., 49, 200 (1972)

25) R. C. Williams and J. W. Taylor, J. Chem. Educ., **47**, 129 (1970)

26) W. C. Childs, J. Chem. Educ., **50**, 290 (1973)

27) D. Cabral, D. Cachet and J. H. Basso, J. Chem. Educ., **52**, 266 (1975)

28) J. J. Marock and K. L. Hooper, J. Chem. Educ., **48**, 530 (1971)

29) A. Foglio Para and E. Lazzarini, J. Chem. Educ., **51**, 336 (1974)

30) P. Empedocles, J. Chem. Educ., **51**, 593 (1974)

31) D. A. Dixon and R. H. Shafer, J. Chem. Educ., **50**, 648 (1973)

32) J. L. Hogg, J. Chem. Educ., **51**, 109 (1974)

33) J. R. Garbarino and M. A. Wartell, J. Chem. Educ., **50**, 792 (1973)

34) J. R. Garbarino and M. A. Wartell, J. Chem. Educ., 51, 484 (1974)

35) A. Streitweiser and P. H. Owens, "Orbital and Electron Density Diagrams. An application of Computer Graphics" McMillan, Riverside, New Jersey, 1973.

36) F. W. Parrett and E. Peterson, J. Chem. Educ., **50**, 122 (1973)

37) D. S. Alderdice and R. S. Watts, J. Chem. Educ., **47**, 123 (1970)

38) D. C. Griffin and J. B. McGhie, Amer. J. Phys., **41**, 1149 (1973)

39) M. Cox and L. R. B. Elton, Amer. J. Phys., **42**, 340 (1974)

40) E. C. Frenkel and D. D. Davis, J. Chem. Educ., **50**, 80 (1973)

41) J. H. Campbell, J. Chem. Educ., **51**, 673 (1974)

42) W. G. Rhodes and L. Brown, J. Chem. Educ., **51**, 595 (1974)

43) J. R. Potts and D. T. Macero, J. Chem. Educ., 50, 494 (1973)

44) W. G. Rhodes and L. Brown, J. Chem. Educ., 51, 595 (1974)

45) W. A. Seitz and F. A. Matson, J. Chem. Educ., 51, 192 (1974)

46) L. W. Richards, J. Chem. Educ., 43, 552 (1966)

47) M. Bader, J. Chem. Educ., 46, 206 (1969)

48) K. A. Mantei and R. L. Hunter, J. Chem. Educ., 51, 213 (1974)

49) F. W. Lima and L. T. Atulla, J. Radioanal. Chem., 20, 769 (1974)

50) J. S. Hinkel, At. Absorpt. Newslett., 13, 127 (1974)

51) E. R. Price, M. J. Stoklosa and J. R. Wasson, J.Chem. Educ., 50, 177 (1973)

52) J. D. Lee, Talanta, 20, 1029 (1973)

53) H. M. Bell, J. Chem. Educ., 51, 548 (1974)

54) B. D. Dombek, J. Lowther and E. Carberry, J. Chem. Educ., 48, 729 (1971)

55) G. Ng, J. Chem. Educ., 52, 91 (1975)

56) M. Bader, J. Chem. Educ., 48, 175 (1971)

57) P. E. Clark and K. D. Berlin, J. Chem. Educ., 49, 362 (1972)

58) R. E. Rondeau and H. A. Rush, J. Chem. Educ., 47, 139 (1970)

59) A. C. Ling, J. Chem. Educ., 51, 174 (1970)

60) K. J. Johnson, J. Chem. Educ., 47, 819 (1970)

61) P. R. Bevington, "Data Reduction in the Physical Sciences", McGraw-Hill, New York, 1969

62) G. Beech, "FORTRAN IV in Chemistry", John Wiley and Sons, Chichester England, 1975

63) A. C. Norris and B. A. Collins, Int. J. Math. Educ. Sci. and Technol., 5, 433 (1974)

64) C. Caso, Amer. J. Phys., 42, 1037 (1974)

65) L. J. Saltzberg, J. Chem. Educ., 49, 357 (1972)

66) G. F. Pollnow, *J. Chem. Educ.*, **48**, 518 (1971)

67) H. Kim, *J. Chem. Educ.*, **47** 120 (1970)

68) J. L. Dye and V. A. Nicely, *J. Chem. Educ.*, **48**, 443 (1971)

69) J. G. Becsey, L. Berke and J. R. Callan, *J. Chem. Educ.*, **45**, 728 (1968)

70) G. Szonyi, *Chem. Technol*, 36 (January 1973)

71) J. W. Connolly, *J. Chem. Educ.*, **49**, 262 (1972)

72) M. A. Wartell and J. A. Hurlbut, *J. Chem. Educ.*, **49**, 508 (1972)

73) C. A. King and A. W. Sangster, *School Science Review*, **51**, 630 (1970)

74) C. B. Leonard, *J. Chem. Educ.*, **47**, 149 (1970)

75) P. D. Groves, *Comput. J.*, **10**, 365 (1968)

76) G. F. Pollnow, *J. Chem. Educ.*, **44**, 679 (1967)

77) J. A. Mann, H. Zeitten and A. B. Delfin, *J. Chem. Educ.*, **44**, 673 (1967)

78) T. A. Gosink, *J. Chem. Educ.*, **48**, 538 (1971)

79) S. P. Perone and J. F. Eagleston, *J. Chem. Educ.*, **48**, 438 (1971)

80) A. Savitzky and M. J. E. Golay, *Anal. Chem.*, **36**, 1627 (1964)

81) W. E. McDermott, Nasa Tech. Memo., 1972 (CNASA TM X-2440) p 744

82) J. L. Spratt, "Liquid Scintillation Counting". Vol.2, edited by M. A. Crook, P. Johnson and B. Scales, Heyden Ltd., London 1971

83) G. Beech, *American Laboratory* p 53 (September 1973) and *International Laboratory*, p 35 (January 1974)

84) B. J. Bulkin, E. H. Cole and A. Noguerola, *J. Chem. Educ.*, **51**, A273 (1974)

85) R. E. Dessy and J. Titus, *Anal. Chem.*, **46**, 294A (1974)

86) S. D. Ward "Mass Spectrometry", (Chemical Society Specialist Reviews) 264 (1973)

# PART FOUR
# Retrospect

## 4.1 Hints and Answers to Selected Problems

In many cases, there are no unique answers because any particular problem can be solved by many routes. Therefore, in many cases, only brief hints are given:

*UNIT 1*
(1) The statements should read:
```
100 LET M=X1 +K      (X10 is illegal - only one digit may follow
                                                         a letter)
120 LET Y=92         (The variable must be on the left of the
                                                         equals)
520 LET R=U+P/(-3)   (Operators must not be adjacent)
999 LET Y=P+Q        (LET should not be omitted - although you
                       can often get away with not using it!)
```

(2) This is a simple modification.

(3)  278           -8.92E1
   0.048376       -3284500
   5.8948873E8    (no more than 8 digits)
   3.72E-11

(4) Rearrange the equation to V=nRT/P and do not use P=0.

*UNIT 2*
(1) Simply retype the line numbers and run the program

(2) Precisely as for (1)

*UNIT 3*
(1)   (i) The DATA line needs one more value to match the READ.
     (ii) Line number missing and no commas between the values.
    (iii) INPUT can not use a pure value such as 8 and the t
          should be upper case, T.

(2) Be careful in the use of brackets;  a statement of the form
         400 LET V=R*(N1↑(-2) - N2↑(-2))
could be used.  Check that N1 or N2 are not entered as zero.
Note that R contains a large number of digits.

(3) Your program could usefully include REM's to indicate the names of the compounds.

*UNIT 4*

(1) 1900 PRINT"MESSAGE"     450 PRINT A$
    900 PRINT Q;"LAST VALUE"

(2) Simple replacement

(3) A suitable PRINT could be
    800 PRINT "PERCENT SULPHUR=";S;"    .CHLORINE";C
    where S and C are the computed values.

(4) Tricky use of brackets can arise. We can calculate $T_2$ from:
    800 LET T2=(P2*V1/(N*R))*(P1/P2)↑G

    where G represents $\gamma$ and the other variables have their obvious meaning.

*UNIT 6*

(1) Input the number of batches B (3 in this case) and use this as an index in an outer loop. Use an inner loop to read the number, N, of analyses of a batch:
    300 FOR I=I TO B
    310 READ N
    320 FOR J=1 TO N
    330 READ P
        ....
    .....
    500 NEXT J
    510 NEXT I

(2) Use two non-nested loops; calculate the mean value firstly and use a RESTORE statement so that you can re-access the original list of results.

*UNIT 7*

(2) A statement of the form:
    100 DIM X(100),Y(100),Z(100),S(2,2)

    could be used - or a separate DIM for S(2,2).

*UNIT 9*

Is best to plot the function in an approximately square 'box' n characters wide and m lines high. The scale in the x and y directions will then be different: for example, if the box has 1 Angstrom sides and if we assume that the centre point x=0, y=0 is at m/2, n/2 then, from the diagram below, we can calculate the true x and y coordinates.

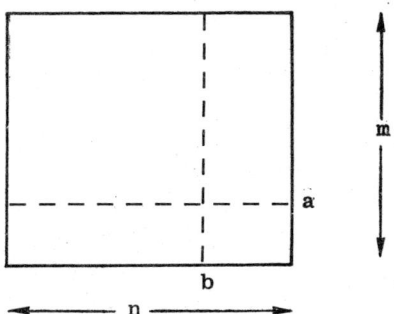

from $\quad x = b - n/2$
$\quad\quad\quad y = a - m/2$

where b is the printer head position across the paper (1<b<n) and a is the line position (1<a<m). The value of r can be calculated from

$$r = \sqrt{x^2 + y^2}$$

and the value of $\Psi$ could be stored in a two-dimensional array such as F(n,m). If the sides of the 'box' are other than 1 Angstrom, then a scale factor is used. We also scale the computed values of $\Psi$ to lie between 1 and 100. Suitable characters can then be printed at each of the (n times m) print positions to represent the value of the function. A suitable choice of characters could be:

```
blank -  0 to 1
    .  -  1 to 3
    ,  -  3 to 5
    "  -  5 to 7
    +  -  7 to 9
    *  -  9 to 10
```

## UNIT 10

(1) Write a subroutine that will calculate any value of a factorial such as m! by using a FOR/NEXT loop in which the value of m! is accumulated as a product.

## UNIT 11

(3) The equations are simply solved by matrix inversion; the solution is $\quad x_1 = 1; x_2 = 2; x_3 = 3; x_4 = 4; x_5 = 5$

## UNIT 12

(2) There are many ways to approach this - for example, the start of a peak can be identified from the point at which the deflection from the baseline exceeds a set value; similarly, the end of a peak can be detected. The peak area is proportional to the sum of the individual deflections. Further details are to be found in the references cited on page 70.

## 4.2 Index/MASTERFILE

The 'page' column contains numbers of the pages on which the particular word or phrase appears; in the 'comment' column write a short definition or description of the word or phrase. Very soon, you will have a useful reference guide which will help you as you write your own programs.

|  | Page | Comment |
|---|---|---|
| ABS | 45 |  |
| Argument | 45 |  |
| Array | 53 |  |
| Assignment | 14 |  |
| Binary | 6 |  |
| Bit | 6 |  |
| Byte | 6 |  |
| BYE | 11,12 |  |
| Compiler | 7 |  |
| CON (in matrix statement) | 55 |  |

|  | Page | Comment |
|---|---|---|
| DATA | 21 | |
| DEF | 46 | |
| DIM | 35, 46 | |
| Dimension | 35 | |
| E format | 9 | |
| END | 15 | |
| Errors, correction of | 8, 12 | |
| File | 61 | |
| Flowchart | 2 | |
| FOR-NEXT loop | 31 | |
| Functions, defined | 46 | |
| Functions, standard | 45 | |
| GOODBYE | 11, 12 | |
| GOSUB | 49 | |
| GOTO | 28 | |

|  | Page | Comment |
|---|---|---|
| High level language | 7 |  |
| IDN (in matrix statement) | 55 |  |
| IF.....THEN | 28 |  |
| INPUT | 21 |  |
| Interactive | 8 |  |
| Inversion (of a matrix) | 56 |  |
| Invitation to type | 10 |  |
| LET | 14 |  |
| Line | 9 |  |
| Line number | 9 |  |
| Loop | 31 |  |
| Low level language | 7 |  |
| MAT | 53 |  |
| MAT PRINT | 54 |  |
| MAT READ | 53 |  |
| Matrix | 53 |  |

|  | Page | Comment |
|---|---|---|
| Nested loop | 32 | |
| NEW | 19, 12 | |
| NEXT | 31 | |
| Non-executable statement | 66 | |
| Numeric constant | 9 | |
| Numeric variable | 14 | |
| OLD | 19, 12 | |
| Output | 4, 5 | |
| PRINT | 14 | |
| Quote marks, use of | 24 | |
| READ | 21 | |
| REM | 22 | |
| RESTORE | 21 | |

| | Page | Comment |
|---|---|---|
| RETURN | 49 | |
| RND | 45 | |
| RUN | 11 | |
| SAVE | 19, 12 | |
| Semicolon in PRINT | 24 | |
| STEP | 31 | |
| String constant | 25 | |
| String variable | 25, 40 | |
| Subroutine | 49 | |
| Subscripted variable | 35 | |
| System command | 10 | |
| TAB | 24 | |
| Teletype | 5 | |
| Terminal | 6 | |
| Time sharing | 6 | |
| Transpose of a matrix | 56 | |

| | Page | Comment |
|---|---|---|
| UNSAVE | 19, 12 | |
| Visual display unit | 6 | |
| Word (binary) | 6 | |
| ZER (in matrix statement) | 55 | |

*To Produce Tone:*
  *Print " press CNTRL and G " (return)*

*System Commands:*
  *System N, "Bye"*
  *System N, "PAU-X"*

*also: LIN(50)*